Carpenters and Builders Library No. 4

by John E. Ball

THEODORE AUDEL & CO.
a division of

HOWARD W. SAMS & CO., INC.
4300 West 62nd Street
Indianapolis, Indiana 46268

FOURTH EDITION

FOURTH PRINTING—1978

International Standard Book Number: 0-672-23243-X
Library of Congress Catalog Card Number: 76-24079

Foreword

Carpentry, as a branch of the building trade, is one of man's oldest and most significant vocations. It encompasses all processes and degrees of construction with wood—from the simple sanding of a board for a cabinet shelf to the complex task of building an entire house. This book, the fourth in a four-book series, provides the knowledge necessary to attain success in any carpentry project attempted by the home owner, hobbyist, apprentice, journeyman, and master carpenter.

The text gives a complete explanation of the numerous and varied power tools which a carpenter might use and provides an intimate understanding of their uses. Power-operated hand tools, boring machines, woodturning lathes, etc., are covered in this book, along with a detailed description of roofing, doors, windows, millwork, and painting, flooring and exterior sheathing are also examined in depth, along with some of the many applications of the varous materials for interior walls and partitions—plaster, fiberboard, plywood, etc.

The other three texts in this series provide a complete and thoroughly rounded range of topics which are of primary concern to both the would-be and the professional woodworker. These topics consist of such carpentry fundamentals as builders mathematics, surveying, architectural drawing, house construction, framing, foundations, how to use the steel square, and the construction of doors, and windows.

The purpose of this volume is to present the methods used to complete the frame structure in its entirety. Knowledge of this and other included subjects is necessary for skilled workmanship in the carpentry and building trades.

Contents

Chapter 1

ROOFING 9

Slope of roof—roll roofing—roofing felt—installation—flashing—canvas roofing—built-up roofs—wood shingles—mineral-surfaced asphalt shingles—cement-asphalt shingles—slate roof—sheet-metal roof—gutters and downspouts—selecting roofing material—detecting leaks—summary—review questions

Chapter 2

CORNICE CONSTRUCTION 45

Box cornices—closed cornices—wide box cornices—open cornices—cornice returns—rake or gable-end finish—summary—review questions

Chapter 3

MITER WORK 53

Miter tools—mouldings—mitering flush moulding—mitering spring mouldings—cutting long miters—coping—summary—review questions

Chapter 4

DOORS 65

Mill-made doors—panel doors—flush doors—solid-core doors—hollow-core doors—louver doors—door frames—door trim—hanging doors—swinging doors—sliding doors—garage doors—summary—review questions

Chapter 5

WINDOWS 85

Window framing—double-hung windows—hinged or casement windows—window sash—glazing—screens—shutters—summary—review questions

Chapter 6

SHEATHING AND SIDING 95

Fiberboard sheathing—wood sheathing—plywood sheathing—sheathing paper—wood siding—bevel siding—drop siding—vertical siding—installation of siding—corner treatment—summary—review questions

Chapter 7

STAIRS 111

Stair construction—ratio of riser to tread—design of stairs— framing stair well—stringer or carriage—basement stair—newels and handrail—exterior stair—disappearing stair—summary—review questions

Chapter 8

FLOORING 129

Subflooring—floor covering—sound-proof floors—wood-tile flooring—linoleum—asphalt—rubber tile—ceramic tile—summary—review questions

Chapter 9

INTERIOR WALLS AND CEILING 143

Lath-plaster reinforcing—installing gypsum base—plaster ground—nailing lath—plaster material and method of application—drywall finish—plywood panels—wood and fiberboard panels—summary—review questions

Chapter 10

CIRCULAR SAWS 165

Construction — tilting arbor — ripping — crosscutting — mitering — grooving — dado head — power and speed of saw.

Chapter 11

BAND SAWS 175

Construction — straight cutting — ripping — circular arcs and segments — multiple sawing — miter gauge and stop rod — miter clamp — operation pointers.

Chapter 12

JIGSAWS 185
Construction — component parts — guides — saw blade — jigsaw cutting — operation.

Chapter 13

WOODTURNING LATHES 191
Lathe speeds—starting and stopping lathe—lathe attachments—measurements—centering and mounting stock—woodturning operations—polishing—summary—review questions

Chapter 14

PLANERS, JOINTERS, AND SHAPERS 207
Planer construction—adjustments—operations—jointer construction—adjustments—operation—shaper—construction—adjustments —operation—summary—review questions

Chapter 15

MORTISERS AND TENONERS 227
Mortisers—tenoners—summary—review questions

Chapter 16

SANDING MACHINES 233
Drum sanding—belt sanding—disk sanding—spindle sanding—block sanding—summary—review questions

Chapter 17

BORING MACHINES 245
Boring tools—summary—review questions

Chapter 18

POWER OPERATED HAND TOOLS 251
Power hand saws—electric drills—bench grinders—electric planes

—saber saws—portable sanders—summary—review questions

Chapter 19

TERMITE PROTECTION 263
    Termite identification—damage—termite shields—detecting termite
location—use of soil poisons—summary—review questions

Chapter 20

PAINTING AND EQUIPMENT 273
    Paints—mixtures—primary colors—interior painting—primer coat
—exterior painting—finishing wood floors—painting surfaces—
painting tools—spray painting—spray gun—compressor—spray-
booth—care of equipment—summary—review questions

Chapter 21

MAINTENANCE AND REPAIR 307
    Maintenance of basement, crawl-space, attic and roof—selection of
lumber—summary—review questions

Chapter 22

PHYSICAL CHARACTERISTICS OF WOOD 319
    Structure of lumber—softwoods and hardwoods—moisture content
—shrinkage—estimating density—strength—summary—review
questions

GLOSSARY OF HOUSING TERMS 331

INDEX 354

CHAPTER 1

Roofing

A roof of a building includes the roof cover (the upper layer which protects against rain, snow, and wind), the sheathing to which it is fastened, and the framing (rafters) which supports the whole structure.

The term roofing, or roof cover, refers to the uppermost part of a roof. Because of its exposure, roofing usually has a limited life, and so is made to be readily replaceable. It may be made of many widely diversified materials, among which are the following:

1. Wood, usually in the form of shingles or shakes.
2. Metal, which may be lock-seamed tin or copper, corrugated steel or aluminum sheets, or any of various ribbed or corrugated patterns.
3. Slate, which may be the natural product , or rigid manufactured slabs often of cement-asbestos.
4. Tile, which is a burned clay or shale product. Several standard types are available. In addition, concrete tiles have been produced, but have not been uniformly satisfactory.
5. Built-up covers of asphalt- or tar-impregnated felts, with moppings of hot tar or asphalt between the plies and a mopping of tar or asphalt over all. With the tar-felt roofs, the top is usually covered with embedded gravel or crushed slag.
6. Roll roofing, as the name implies, is marketed in rolls containing approximately 108 square feet. Each roll is usually

36 inches wide, and may be plain or have a coating of various colored mineral granules. The base is a heavy asphalt-impregnated felt.

7. Asphalt-felt flexible shingles, usually in the form of strips with two, three, or four tabs per unit. These shingles are asphalt, and with the surface exposed to the weather heavily coated with mineral granules. Because of their fire-resistance, nominal cost, and reasonably good durability, this is the most popular roofing material for residences at the present time. They are available in a wide range of colors, including black and white.

8. Synthetic plastics. This includes the corrugated, glass-fibered sheets, which are obtainable in various colors. They are translucent, with the percentage of solar radiation passing through depending greatly on their color. They are used for skylighting, and for patio and marquee roofs. Synthetic rubber coatings are available which may be applied over sheathings of plywood, canvas or roll roofing, or concrete. They are available in various colors, are highly elastic, and will not break at seams or joints. This type of roofing material is ideally suited for heavy foot traffic on sun decks.

9. Canvas is a very old and extremely satisfactory roof and deck cover used for sun decks and other locations subject to heavy foot traffic. It has been used for many years on the decks of boats and ships.

SLOPE OF ROOFS

The slope of the roof is frequently a factor in the choice of roofing materials and in the method used to put them in place. The lower the pitch of the roof the greater is the chance of wind getting under the shingles and tearing them out. Interlocking asbestos-cement shingles and cedar shingles resist this wind prying much better than the standard asphalt shingles. For very low-pitched slopes, the manufacturers of asphalt shingles recommend that the roof be planned for some other type of covering.

Asphalt shingles can be used satisfactorily, however, on moderate sloping roofs if the heavy butt shingles are used. Most of the manufacturers make a seal-down type, which in a short time will bond with the shingle below. Metal shingles and aluminum strip roofing virtually eliminate the problem of wind prying, but they are noisy. Most homeowners object to the noise during a rain storm. Even on porches, this noise is often annoying inside the house.

Spaced roofing boards are frequently used with cedar shingles as an economy measure, and also because the cedar shingles themselves add considerably to the strength of the roof. The spaced roofing boards, however, reduce the insulating qualities, and it is advisable to use a tightly sheathed roof beneath the shingles if the need for insulation overcomes the need for economy. Spaced roofing boards are not satisfactory for asphalt shingles.

Because of drainage consideration, most roofs should have a certain amount of slope. Roofs which are covered with tar-and-gravel coverings are quite satisfactory when built level. Such covers are entirely watertight, and water does not harm the material. Some shallow puddles of water invariably will stand just after a rain, but they can do no harm, and evaporate quickly. Level roofs drain very slowly, and slightly smaller eaves troughs and down spouts are used on these roofs. They are quite common on industrial and commercial buildlings.

ROLL ROOFING

Roll roofing is an economical cover especially suited for roofs with low pitches. It has a base of heavy asphalt-impregnated felt with additional coatings of asphalt that are dusted to prevent adhesion in the roll. The weather surface may be plain, or covered with fine mineral granules. Many different colors are available. One edge of the sheet is left plain (no granules) where the lap cement is applied. For best results, the sheathing must be tight, preferably 1 × 6 tongue-and-groove, or plywood. If the sheathing is smooth, with no cupped boards or other protuberance, the slate-surfaced roll roofings will withstand a surprising amount of abrasion from foot traffic, although it is not generally recommended for that purpose. Windstorms are the most relentless

11

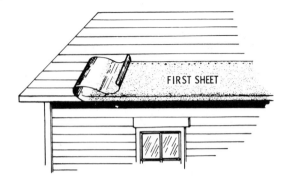

Fig. 1. Applying the first strip of roll roofing.

enemy of roll roofings. If the wind gets under a loose edge, almost certainly a section will be blown off.

Roofing Felt

Roll roofing can be applied directly to the wood sheathing, or can be installed over roofing felt. The roofing felt is relatively inexpensive and its use is suggested when a first-class roofing job is desired. The felt is available in two weights or thicknesses— 15 lb. and 30 lb. The 15-lb. weight is normally used as a base.

To apply the felt, start the first strip at the eave, allowing it to project 1/2 inch. Hold this strip in place with a few nails or tack it in place with a staple gun. Apply the second strip, allow-it to overlap the first by about 2 inches. Tack this in place also. The entire roof is covered in this manner, after which the roll roofing is applied.

Care must be taken if the day is very windy to keep the felt from being ripped from the roof. Under these conditions, it is better to apply a strip of felt and then a strip of roofing instead of first covering the entire roof area with felt.

Installation

The first strip of roll roofing is applied along the eave, with from 1 to 2 inches projecting over the eave. The strip should be cut to length on the ground to make it easier to handle. Measure the exact length of the roof from edge to edge and add 1 3/8 inches for each end to extend over the gable ends. Make sure the ends are cut square. If they are not, allow enough extra projection so that a final projection of 1 3/8 inches will be left when the squaring of the cut is made.

If the eave is uneven, or if it is difficult to position the first strip in a straight line, a chalk line struck across the roof will be useful. Measure up the roof from the eave along each gable end and mark the correct distance. Strike a chalk line between these two marks. Now the top edge of the first strip of roofing can be aligned with this chalked line. The same procedure can be used for each succeeding strip of roofing.

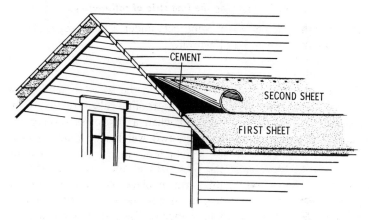

Fig. 2. Method of cementing and lapping the first and second strips of roll roofing.

As shown in Fig. 1, nail along the top edge of the first strip, spacing the nails from 12 to 16 inches apart and about 1 inch from the top edge. Apply roofing cement along the top 2 inches of the strip and place the second piece of roll roofing in position (Fig. 2). Let the second strip overlap the first by 2 inches (Fig. 3).

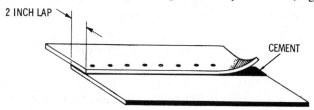

Fig. 3. Width of lap and area to be cemented.

Nail the roofing in place along the overlap with the correct length galvanized roofing nails spaced 2 inches apart and 1 inch from the edge of the roofing. Continue this procedure until the roofing reaches the ridge. Cut the strip, if necessary, to lie even along the ridge. Cover the other side of the roof in the same way.

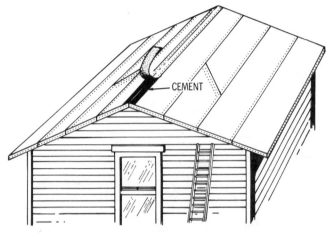

Fig. 4. Completed roof with roofing ridge being installed. Note method of roof nailing.

Finally, cut a strip of roofing from 9 to 12 inches wide for the ridge cap. Cement and nail this strip in place, as shown in Fig. 4.

The edges of the roofing that project over the eaves and gable ends should be turned down and securely nailed until they look similar to the roof in Fig. 5. It is important that the edge of the turned-down roofing project past the wood to which it is nailed by at least 3/8 inch to permit water to drip off and not run down on the wood finish. This will prevent discoloration of the painted surface and eventual destruction of the wood.

To save material, the roof strips can be spliced when they do not reach the full length of the roof. The splicing method shown in Fig. 5 can be used. Cut each sheet at an angle where they are to be joined, allowing a 4-inch overlap. Cement the angled edge of the bottom strip and nail the top strip in place with two rows of nails. The angle cut permits the water to run off the splice instead of along it, which it would do if the two pieces were cut at right angles.

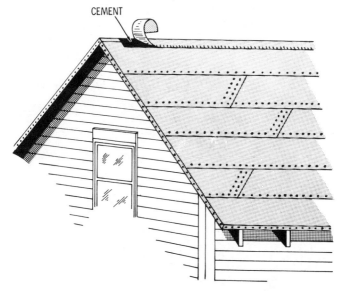

CEMENT

Fig. 5. Method of splicing sheets of roll roofing.

When applying roofing on a roof that is too steep to work on without slipping, the method illustrated in Fig. 6 can be used. The first sheet of roofing is laid along the ridge, with a few nails along the top edge to hold it in place. Nail a 2 × 4 over the roof sheathing, 30 inches from the lower edge of the first 2 × 4. These are to stand or kneel on, and if they are moved down as the work progresses, the roofing will not have to be walked on.

Unroll the second sheet of roofing, allowing it to rest against the 2 × 4, its upper edge temporarily overlapping the lower edge of the first sheet by 2 inches (Fig. 6A). Lift the upper edge of the second sheet and place it under the lower edge of the first sheet, thus reversing their positions and making the lap so it will shed water. Lift the lower edge of the first sheet and apply cement to the upper edge of the second sheet. Begin nailing in the middle of the roof and work out to each end.

Remove the upper 2 × 4 and nail it 30 inches below the lower 2 × 4. Apply the third sheet of roofing in the same manner as the

15

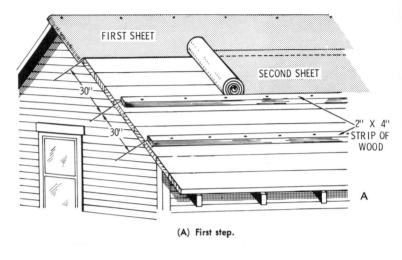

(A) First step.

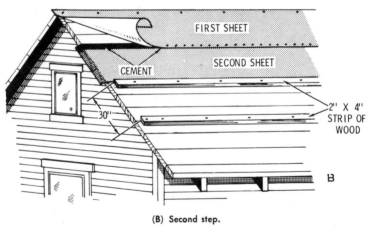

(B) Second step.

Fig. 6. Method of applying roll roofing on a steep roof.

second. Continue this process down the roof. Finish the job by applying the ridge and nailing down the projecting edges.

Valleys

For a roof that contains valleys, the method shown in Fig. 7 should be used. A valley strip is cut wide enough to extend under the roofing for a distance of at least 6 inches. This strip is installed first, after which the sheets of roll roofing are applied. The valley ends of the sheets are cut at an angle, cement is

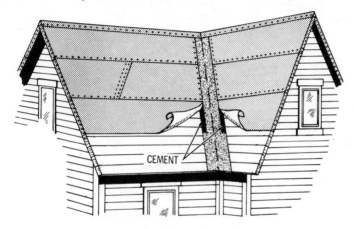

Fig. 7. Method of laying valleys.

applied under them, and they are nailed as shown. *Do not nail through any exposed portion of the valley.*

Flashing

When a roof abuts against a vertical wall, flashing can be made by extending the roof up the vertical wall for a distance of 4 inches, as in Fig. 8. Fit the roofing snugly into the angle, but do not break it. Cement the turned-up part to the wall. Cut a strip of roofing 12 inches wide and fit it over the turned-up edge, as shown. Cement the upper part of this strip to the wall so no water can get behind it. Cement the lower part where it overlaps the turned-up roofing. Fasten it in place with a strip of wood having its upper edge beveled.

For a brick wall, scrape the mortar out of a level joint to a depth of about 1 inch. Turn the upper edge of the turned-up roofing into this mortar joint and pack with Portland cement. Raggle or flashing blocks (Fig. 9) laid as part of the vertical wall provide a convenient method of flashing. The roofing is turned up the wall and cemented to the curved notch of the raggle block.

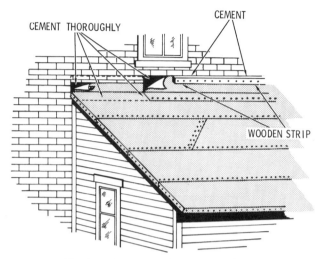

Fig. 8. Joining a roof to a vertical wall.

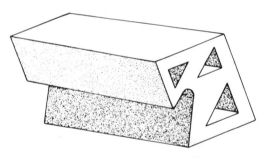

Fig. 9. A raggle or flashing block, 4" x 5" x 8".
This block can be built into masonry walls to
provide effective flashing construction.

Metal flashings are always preferable and should be used when possible. The same procedure for cementing and nailing is used for metal as for other types of flashing.

Flashing around chimneys and skylights is installed in the same general manner as for vertical walls. Fig. 10 shows the method of applying the flashing, whether it is metal or roofing material. The generous use of cement and nails, along with overlaps of 6 inches or more, will result in a watertight joint. The pattern and method of cutting and folding a chimney flashing is shown in Fig. 11.

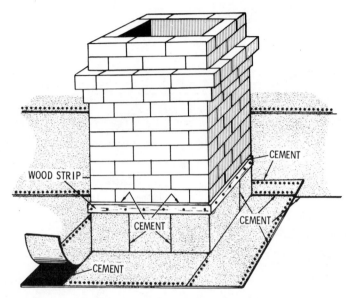

Fig. 10. Method of flashing a chimney.

CANVAS ROOFING

Canvas is often used for waterproofing boat decks and for sun decks subjected to foot traffic. Canvas (or cotton duck) suitable for roofs or decks is made in two general classes. The highest number in the "numbered" ducks is 12, which designates the lightest weight, weighing 7 ozs. per yard, 22 inches wide. The lowest number is 00, which weights 20 ozs. per yard. "Ounce" ducks weigh from 6 to 16 ozs., per yard, 28 1/2 inches wide. The grade called "army" is generally used for roofing, with the 10-oz. weight being the lightest weight recommended.

The sheathing for this type of covering must be smooth. The canvas will quickly wear through at the turned-up edges of cupped boards. Tongue-and-grooved 1 × 4 No. 2 flooring makes a good deck.

19

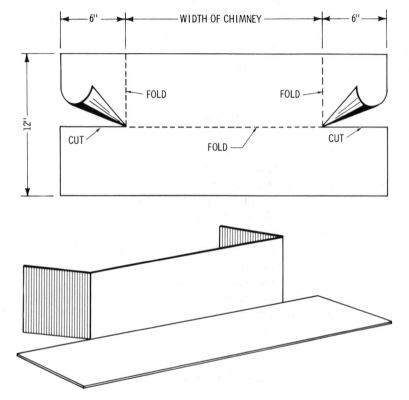

Fig. 11. Pattern for cutting chimney flashing.

The deck (sheathing) is first primed with a mixture made up in the proportions of 100 lbs. of heavy paste white lead in oil, 4 gallons of raw linseed oil, 2 gallons of turpentine, and 1 pint of japan drier. Smaller quantities may be mixed in these proportions. Apply the primer and allow it to dry. Then give the deck a *heavy* coat of the heavy paste white lead. Immediately lay the canvas over this lead-coated surface and roll with a linoleum roller. The edges should be lapped 1 1/2 inches, and leaded between the laps. Nail the laps with 3/4-in. hard copper tacks, spaced 3/4 inch apart.

After the canvas has been applied and rolled, it should be primed with a mixture made up in the following: The proportions of 100 lbs. heavy paste white lead in oil, 3 gallons raw linseed oil, 2 gallons turpentine, and 1 pint japan drier. After the primer has dried, the deck should be given two coats of a good exterior oil-based paint. Decks installed in this manner should give good service under moderate foot traffic for 25 to 30 years, or more, if kept well painted.

BUILT-UP ROOF

A built-up roof is constructed of: sheathing paper; a bonded base sheet; perforated felt; asphalt; and surface aggregates. Fig. 12. The sheathing paper comes in 36 inch wide rolls and has approximately 500 square feet per roll. It is a rosin-size paper and is used to prevent asphalt leakage to the wood deck. The base sheet is a heavy asphalt saturated felt that is placed over the sheathing paper. It is available in 1, 1½, and 2 square rolls. The perforated felt is one of the primary parts of a built-up roof. It is saturated with asphalt and has tiny perforations throughout the sheet. The perforations prevent air entrapment between the layers of felt. The perforated felt is 36 inches wide and weighs approximately 15 pounds per square. Asphalt is also one of the basic ingredients of a built-up roof. There are many different grades of asphalt, but the most common are; low melt, medium melt, high melt, and extra high melt.

Prior to the application of the built-up roof, the deck should be inspected for soundness. Wood board decks should be constructed

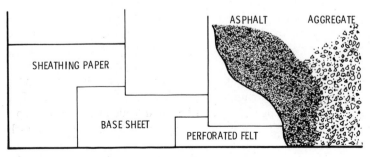

Fig. 12. Sectional plan of a built-up roof.

21

of ¾ inch seasoned lumber. Any knot holes larger than one inch should be covered with sheet metal. If plywood is used as a roof deck it should be placed at right angles to the rafters and be at least ½ inch in thickness.

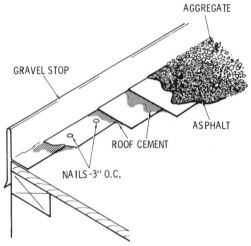

Fig. 13. Illustrating the gravel stop.

The first step in the application of a built-up roof is the placing of sheathing paper and base sheet. The sheathing paper should be lapped 2 inches and secured with just enough nails to hold it in place. The base sheet is then placed with 2 inch side laps and 6 inch end laps. The base sheet should be secured with ½"-diameter head galvanized roofing nails placed 12 inches on center on the exposed lap. Nails should also be placed down the center of the base sheet. The nails should be placed in two parallel rows 12 inches apart.

The base sheet is then coated with a uniform layer of hot asphalt. While the asphalt is still hot, layers of roofing felt are placed. Each sheet is then coated with a uniform layer of hot asphalt. While the asphalt is still hot, layers of roofing felt are placed. Each sheet should be lapped 19 inches, leaving an exposed lap of 17 inches.

Once the roofing felt is placed, a gravel stop is installed around the deck perimeter, Fig. 13. Two coated layers of felt should extend 6 inches past the roof decking where the gravel stop is to be installed. When the other piles are placed, the first two layers are folded over the other layers and mopped in place. The gravel stop is then placed in a ⅛ inch thick bed of flashing cement and securely nailed every 6 inches. The ends of the gravel stop should be lapped 6 inches and packed in flashing cement.

After the gravel stop is placed, the roof is flooded with hot asphalt and the surface aggregate is embedded in the flood coat. The aggregates should be hard, dry, opaque, and free of any dust or foreign matter. The size of the aggregates should range from ¼ inch to ⅝ inch. When the aggregate is piled on the roof it should be placed on a spot that has been mopped with asphalt. This technique assures proper adhesion in all areas of the roof.

WOOD SHINGLES

The popularity of wood shingles for roofing has decreased drastically in the past few years. The chief reason for this decline has been the introduction of improved types of other roof coverings that are not as combustible as wood shingles. Most cities and incorporated towns now prohibit the use of wood shingles because of the potential fire hazard they present. For the same reason, many insurance companies will either refuse to insure a building having a roof of wood shingles, or will charge a higher rate.

In spite of the greater fire risk, wood shingles are still used on many buildings. The better grades of shingles are made of cypress, cedar, and redwood and are available in lengths of 16 and 18 inches and thicknesses at the butt of ⁵⁄₁₆″ and ⁷⁄₁₆″, respectively. They are packaged in bundles of approximately 200 shingles in random width from 3 to 12 inches.

An important requirement in applying wood shingles is that *each shingle should lap over the two courses below it, so that there will always be at least three layers of shingles at every point on the roof.* This requires that the amount of shingle exposed to the weather (the spacing of the courses) should be less than 1/3 the length of the shingle. Thus in Fig. 14, 5½ inches is the maximum amount that 18-inch shingles can be laid to the weather and have an adequate amount of lap. This is further shown in Fig. 15A.

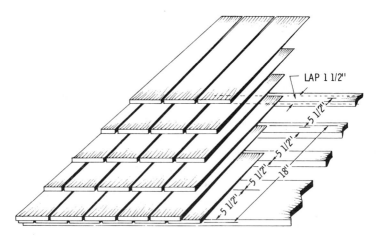

Fig. 14. Section of a shingle roof illustrating the amount of shingle which may be exposed to the weather as governed by the lap.

In case the shingles are laid more than 1/3 of their length to the weather, there will be a space, as shown by MS in Fig. 15B, where only two layers of shingles will cover the roof. This is objectionable, because if the top shingle splits above the edge of the shingle below, water will leak through. The maximum spacing to the weather for 16-inch shingles should be 4-7/8 inches, and for 18-inch shingles should be 5-1/2 inches. Strictly speaking, the amount of lap should be governed by the pitch of the roof. The maximum spacing may be followed for roofs of moderate pitch, but for roofs of small pitch, more lap should be allowed, and for a steep pitch the lap may be reduced somewhat, but it is not advisable to do so. Wood shingles should not be used on pitches less than 4 in. per foot.

Table 1 shows the number of square feet that 1,000 (five bundles) shingles will cover for various exposures. This table does not allow for waste on hip and valley roofs.

Shingles should not be laid too close together, for they will swell when wet, causing them to bulge and split. Seasoned shingles should not be laid with their edges nearer than 3/16 inches when laid by the American method. It is advisable to thoroughly soak the bundles before opening.

Great care must be used in nailing wide shingles. When they

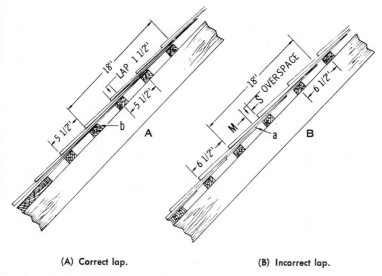

(A) Correct lap.   (B) Incorrect lap.

Fig. 15. The amount of lap is an important factor in applying wood shingles.

Table 1. Space Covered by 1,000 Shingles

Exposure to weather	4¼	4½	4¾	5	5½	6
Area covered in sq. ft.	118	125	131	138	152	166

are over 8 inches in width, they should be split and laid as two shingles. The nails should be spaced such that the space between them is as small as is practical, thus directing the contraction and expansion of the shingle toward the edges. This lessens the danger of wide shingles splitting in or near the center and over joints beneath. Shingling is always started from the bottom and laid from the eaves or cornice up.

There are various methods of laying shingles, the most common known as:

1. The straight-edge.
2. The chalk-line.
3. The gauge-and-hatchet.

The straight-edge method is one of the oldest. A straight-edge having a width equal to the spacing to the weather or the distance

25

between courses is used. This eliminates measuring, it being necessary only to keep the lower edge flush with the lower edge of the course of shingles just laid; the upper edge of the straight edge is then in line for the next course. This is considered to be the slowest of the three methods.

The chalk-line method consists in snapping a chalk line for each course. To save time, two or three lines may be snapped

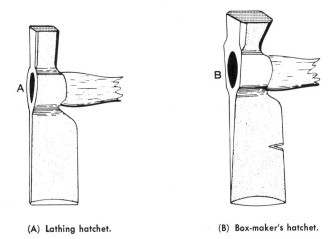

(A) Lathing hatchet.   (B) Box-maker's hatchet.

Fig. 16. Hatchets used in shingling.

at the same time, making it possible to carry two or three courses at once. This method is still extensively used. It is faster than the straight-edge method, but not as fast as the gauge-and-hatchet method.

The gauge-and-hatchet method is extensively used in the Western states. The hatchet used is either a lathing or a box-maker's hatchet, as shown in Fig. 16. Hatchet gauges to measure the space between courses are shown in Fig. 17. The gauge is set on the blade at a distance from the hatchet poll equal to the exposure desired for the shingles.

Nail as close to the butts as possible, if the nails will be well-covered by the next course. Only galvanized shingle nails should be used. The 3d shingle nail is slightly larger in diameter than the 3d common nail, and has a slightly larger head.

26

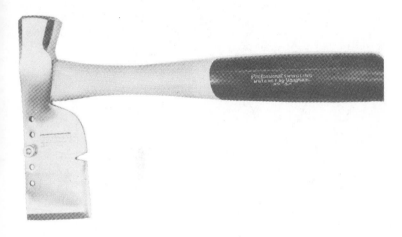

Fig. 17. Shingling hatchet.

Hips

The hip is less liable to leak than any other part of the roof as the water runs away from it. However, since it is so prominent, the work should be well done. Fig. 18 shows the method of cutting shingle butts for hip roof. After the courses 1 and 2 are laid, the top corners over the hip are trimmed off with a sharp shingling hatchet kept keen for that purpose and shingle 3 with the butt cut so as to continue the straight line of courses and again on the dotted line 4, so that shingle *a,* of the second course squares against it and so on from side to side, each alternately lapping the other at the hip joint. When gables are shingled, this same method may be used up the rake of the roof if the pitch is moderate to steep. It cannot be effectively used with flat pitches. The shingles used should be ripped to uniform width.

For best construction, tin shingles should be laid under the hip shingles, as shown in Fig. 19. These tin shingles should correspond in shape to that of hip shingles. They should be at least 7 inches wide and large enough to reach well under the tin shingles of the course above, as at *w.* At *a,* the tin shingles are

27

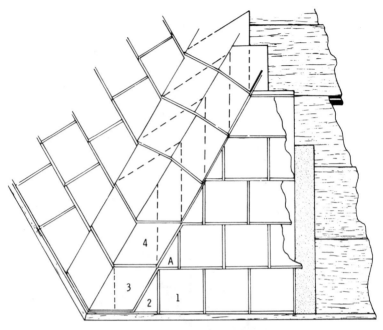

Fig. 18. Hip roof shingling.

laid so that the lower end will just be covered by the hip shingle of the course above.

Valleys

In shingling a valley, first a strip of tin, lead, zinc, or copper, ordinarily 20 inches wide, is laid in the valley. Fig. 20 illustrates an open type valley. Here the dotted lines show the tin or other material used as flashing under the shingles. If the pitch is above 30°, then a width of 16 inches is sufficient; if flatter, the width should be more. In a long valley, its width between shingles should increase in width from top to bottom about 1 inch, and at the top 2 inches is ample width. This is to prevent ice or other objects from wedging when slipping down. The shingles taper to the butt, the reverse of the hip, and need no reinforcing, as the thin edge is held and protected from splitting off by the shingle above it. Care must always be taken to nail the shingle nearest the valley as far from it as practical by placing the nail higher up.

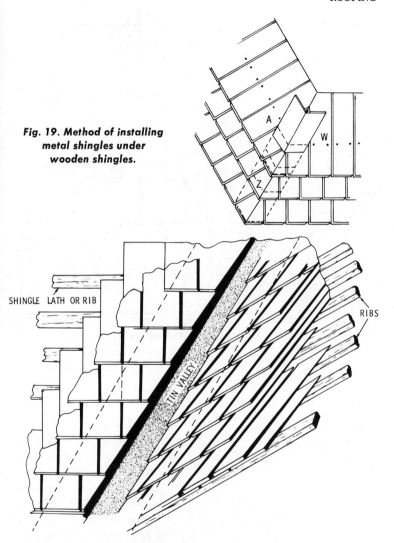

Fig. 19. Method of installing
metal shingles under
wooden shingles.

SHINGLE LATH OR RIB

RIBS

TIN VALLEY

Fig. 20. Method of shingling a valley.

ASPHALT SHINGLES

Asphalt shingles are made in strips of two, three or four units
or tabs joined together, as well as in the form of individual shingles.

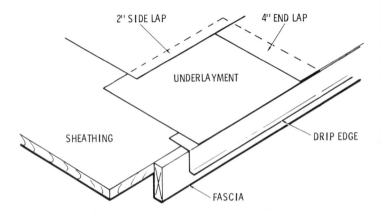

Fig. 21. Application of the underlayment.

When laid, strip shingles furnish practically the same pattern as undivided shingles. Both strip and individual types are available in different shapes, sizes, and colors to suit various requirements.

Asphalt shingles must be applied on slopes having an incline of four inches or more to the foot. Before the shingles are laid, the underlayment should be placed. The underlayment should be 15 lb. asphalt-saturated felt. The underlayment should be placed with 2 inch side laps and 4 inch end laps. Fig. 21. The underlayment serves three purposes: (1) it acts as a primary barrier against moisture penetration, (2) it acts as a secondary barrier against moisture penetration, and (3) it acts as a buffer between the resinous areas of the decking and the asphalt shingles. A heavy felt should not be used as underlayment. The heavy felt would act as a vapor barrier and would permit the accumulation of moisture between the underlayment and the roof deck.

The roof deck should be constructed of well seasoned 1 × 6-inch tongue-and-grooved sheathing. The boards should be secured with two 8d nails in each rafter. Plywood roof sheathing can also be used. It should be placed with the long dimension perpendicular to the rafters. The plywood should never be less than ⅜ inch thick.

To efficiently shed water at the roofs edge, a drip edge is usually installed. A drip edge is constructed of corrosion-resistant sheet metal, and extends 3 inches back from the roof edge. To form the drip-edge the sheet metal is bent down over the roof edges.

The nails used to apply asphalt shingles should be hot-galvanized nails, with large heads, sharp points and barbed shanks. The nails should be long enough to penetrate the roof decking at least ¾ of an inch.

To insure proper shingle alignment, horizontal and vertical chalk lines should be placed on the underlayment. It is usually recommended that the lines be placed 10 or 20 inches apart. The first course of shingles placed is the starter course. The starter course is used to back up the first regular course of shingles and to fill in the spaces between the tabs. The starter course is placed with the tabs facing up the roof and is allowed to project one inch over the rake and eave, Fig. 22. To insure that all cutouts are covered 3 inches should be cut off the first starter shingle.

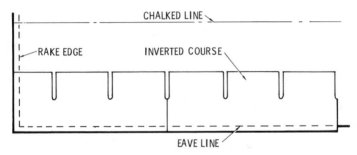

Fig. 22. The starter course.

Once the starter course has been placed, the different courses of shingles can be laid. The first regular course of shingles should be

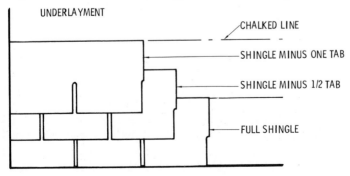

Fig. 23. Application of the starter shingles.

started with a full shingle; the second course with a full shingle, minus ½ a tab; the third course is started with a full shingle; Fig. 23, and the process is repeated. As the shingles are placed, they should be properly nailed. Fig. 24. If a three tab shingle is used, a minimum of 4 nails per strip should be used. The nails should be placed 5⅝ inches from the bottom of the shingle and should be located over the cutouts. The nails on each end of shingle should be located one inch from the end. The nails should be driven straight and flush with the surface of the shingle.

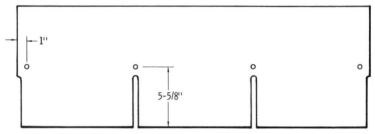

Fig. 24. The proper placement of nails.

If there is a valley in the roof, it must be properly flashed. The two materials that are most often used for valley flashing is 90 lb. mineral surfaced asphalt roll roofing or galvanized sheet metal. The flashing is 18 inches in width and should extend the full length of the valley. Before the shingles are laid to the valley chalked lines are placed along the valley. The chalk lines should be 6 inches apart at the top of the valley and should widen ⅛ inch per foot as as they approach the eave line. The shingles are laid up to the chalked lines and trimmed to fit.

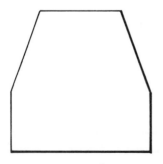

Fig. 25. Hip shingles.

Hips and ridges are finished by using manufactured hip and ridge units, or hip and ridge units cut from a strip shingle. If the unit is cut from a strip shingle, the two cut lines should be cut at an angle. Fig. 25. This will prevent the projection of the shingle past the overlaid shingle. Each shingle should be bent down the center so that there is an equal distance on each side. In cold weather the shingle should be warmed before they are bent. Starting at the bottom of the hip or at the end of a ridge the shingles are placed with a 5 inch exposure. To secure the shingles, a nail is placed on each side of the shingle. The nails should be placed 5½ inches back from the exposed edge and one inch up from the side.

CEMENT-ASBESTOS SHINGLES

These are manufactured from Portland cement and asbestos fiber formed in molds under high pressure. The finished product is hard, fairly tough, and durable. Asbestos-cement shingles are available in a variety of colors and textures and may be obtained in rectangular, square, and hexagonal shape. The use of water-proof felt under shingles is necessary, as indicated in Fig. 21. Lay each shingle with a 2-inch head lap and secure it with slating nails. Drive the nails so that their heads just touch the shingles. Bed all shingles on each side of hips and ridge within one foot of the top and along gable rakes within one foot of the edge in elastic slater's cement. Project the shingles 1 inch over the rear edges of the gutters.

Asbestos-cement shingles may be applied over an old roof covering provided it is in reasonably good condition. The framing should be inspected and, if necessary, reinforced to carry the additional weight of the new shingles with safety. Where the new roofing is to be laid over the old wood shingles, loose shingles should be securely nailed, and warped, split or decayed shingles replaced. Usually the amount of exposure to the weather for the new shingles will not be the same as for the old. To make the new shingles lay flat, beveled strips $3/8 \times 4$ inches should be applied against the butt ends of the old shingles. Wood lath laid end to end may also be used for this purpose.

If the old roofing is in poor condition, it may be more economical to remove it entirely than to make the repairs necessary to provide a sound, smooth surface for the new roofing. If the old roofing is removed, loose sheathing boards should be securely nailed and defective material replaced. If there is openings between the old sheathing boards, they should be filled with new boards of the same thickness as the existing sheathing.

SLATE

Slate is an ideal roofing material and is used on permanent buildings with pitched roofs. The process of manufacture is to split the quarried slate blocks horizontally to a suitable thickness, and to cut vertically to the approximate sizes required. The slates are then passed through planers, and after the operation are ready to be reduced to the exact dimensions on rubbing beds or through the use of air tools and other special machinery.

Roofing slate is usually available in various colors and in standard sizes suitable for the most exacting requirements. On all boarding to be covered with slate, asphalt-saturated rag felt of certain specified thickness is required. This felt should be laid in a horizontal layer with joints lapped toward the eaves and at the ends at least 2 inches. A well secured lap at the end is necessary to properly hold the felt in place, and to protect the structure until covered by the slate. In laying the slate, the entire surface of all main and porch roofs shall be covered with slate in a proper and water-tight manner.

The slate shall project 2 inches at the eaves and 1 inch at all gable ends, and shall be laid in horizontal courses with the standard 3-inch headlap. Each course shall break joints with the preceding one. Slates at the eaves or cornice line shall be doubled and canted 1/4 inch by a wooden cant strip. Slates overlapping sheet-metal work shall have the nails so placed as to avoid puncturing the sheet metal. Exposed nails shall be permissible only in courses where unavoidable. Neatly fit the slate around any pipes, ventilators, etc.

Nails shall not be driven in so far as to produce a strain on the slate. Cover all exposed nail heads with elastic cement. Hip slates and ridge slates shall be laid in elastic cement spread thickly over unexposed surfaces. Build in and place all flashing pieces furnished by the sheeting contractor and cooperate with him in doing the work of flashing. On completion, all slate must be sound, whole, and clean, and the roof shall be left in every respect tight and a neat example of workmanship.

The most frequently needed repair of slate roofs is the replacement of broken slates. When such replacements are necessary, supports similar to those shown in Fig. 26 should be placed

Fig. 26. Illustrating two types of supports used in repairs of roof.

on the roof to distribute the weight of the roofers while they are working. Broken slates should be removed by cutting or drawing out the nails with a ripper tool. A new slate shingle of the same color and size as the old should be inserted and fastened by nailing through the vertical joint of the slates in the overlying course approximately 2 inches below the butt of the slate in the second course, as shown in Fig. 27.

Fig. 27. Method of inserting new pieces of slate shingles.

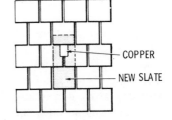

A piece of sheet copper or terneplate about 3 × 8 inches should be inserted over the nail head to extend about 2 inches under the second course above the replaced shingle. The metal strip should

35

be bent slightly before being inserted so that it will stay securely in place. Very old slate roofs sometimes fail because the nails used to fasten the slates have rusted. In such cases, the entire roof covering should be removed and replaced, including the felt underlay materials. The sheathing and rafters should be examined and any broken boards replaced with new material. All loose board should be nailed in place and, before laying the felt, the sheathing should be swept clean, protruding nails driven in, and any rough edges trimmed smooth.

If the former roof was slate, all slates that are still in good condition may be salvaged and relaid. New slates should be the same size as the old ones and should match the original slates as nearly as possible in color and texture. The area to be covered should govern the size of slates to be used and whatever the size, the slates may be of random widths, but they should be of uniform length and punched for a head lap of not less than 3 inches. The roof slates should be laid with a 3-inch head lap and fastened with two large-head slating nails. Nails should not be driven too tight, the nail heads barely touch the slate. All slates within 1 foot of the top and along the gable rakes of the roof should be bedded in flashing cement.

SHEET-METAL ROOFING

Sheet metal used in roof coverings are galvanized iron, tin, aluminum, copper, zinc, and monel metal. These roofing materials are comparatively light in weight and, if properly laid, will provide a watertight and lasting roof. Some metal roofings require the service of professional roofers for application, whereas others, such as corrugated sheets, V-crimp, and pressed standing-seam roofings, are not too difficult for the average home owner to apply, as shown in Fig. 28.

Standing seam construction is the most common method of application when copper is used as roofing material. This construction gives an attractive appearance, and allows for expansion and contraction, since there are no solder joints. Standing-seam construction is suitable for buildings of all types having a roof slope of 2-1/2 inches or more per foot. Standing seams are usually made to finish 1 inch high. Economical spacing of seam calls

**Fig. 28. Showing method of laying corrugated sheet-
metal roof without the use of sheathing.**

for copper sheets in stock widths of 20 inches. Standing seams should never be riveted or soldered. Cross seams need not be soldered if the pitch of the roof is sufficient.

Leaks sometimes develop from improper installation of the metal or failure of the nailing. Such leaks may be corrected by renailing or replacing all or part of the defective sheets. Leaks resulting from faulty joints may be repaired without replacing the metal. Soldered flat seams and joints may be resoldered. Standing seams may be reformed and caulked. Roofings of terne (roofing tin) should be painted at regular intervals. The need for painting will vary in different locations and under different atmospheric conditions, but should not be deferred until rust appears. The surface should be carefully brushed and cleaned of all foreign matter before painting, and rust should be removed as completely as possible with a wire brush.

Two coats of iron-oxide paint are suitable for use on terneplate roofs, but this paint will not usually adhere to new galvanized metal. If painting seems desirable, a chemical treatment with crystalline zinc phosphate should first be used. This may be followed with any good paint suitable for iron or steel. Galvanized steel should not be acid etched before painting. It removes some of the protective zinc coating.

Although metal roofings may be applied over old roofs as well as to new roof decks, it is usually considered better practice to remove the old roofing. If the old roofing is removed, the deck should be repaired so that it will have a smooth, solid surface.

GUTTERS AND DOWNSPOUTS

Most roofs require gutters and downspouts in order to convey the water to the sewer or outlet. They are usually built of metal. In regions of heavy snow fall, the outer edge of the gutter should be ½ inch below the extended slope of the roof to prevent snow banking on the edge of the roof and causing leaks. The hanging gutter is best adapted to such construction.

Downspouts should be large enough to remove the water from the gutters. A common fault is to make the gutter outlet the same size as the downspout. At 18 inches below the gutter, a downspout has nearly four times the water carrying capacity of the inlet at the gutter. Therefore, an ample entrance to the downspout should be provided. Conductor heads or funnels are readily available from roofing establishments and should be used where branch downspouts converge or at the scuppers of flat roofs. Wire baskets or guards should be placed at gutter outlets to prevent leaves and trash from collecting in the downspouts and causing damage during freezing weather.

In cold climates where water will freeze if it should stand in the downspouts, the use of corrugated instead of plain metal will save much trouble and probably prevent the pipes from bursting because of expansion. Various types of downspouts and fittings are shown in Fig. 29. Downspout pipes are usually prefabricated in sections which fit into one another. To insert a length of pipe, the upper section should be slipped into the lower so that water will flow on the inside and not leak out.

Sometimes the joints are soldered tight, but for general practice this is not necessary. Downspouts should be soldered to the straps that fasten to the building. The lower end should have a shoe or turn-out when the water is to be wasted on well-drained ground. A cast-iron pipe connection or boot should be installed when water is to be diverted into a storm sewer, or with a rain switch or diverter to a cistern. Intense rains occur periodically

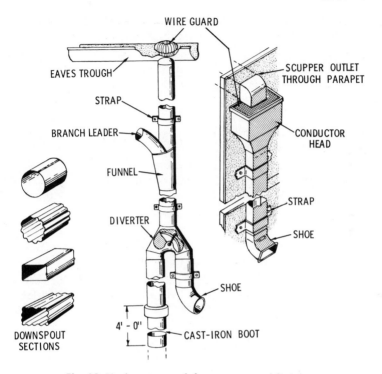

Fig. 29. Various types of downspouts and fittings.

in certain localities but do little harm to the contents of surrounding structures if the gutters overflow for the duration of the storm. For the sake of economy, gutter and downspouts will be ample in size if large enough to carry off only normal storm-water flow.

SELECTING ROOFING MATERIALS

Roofing materials are commonly sold by dealers or manufacturers on the basis of quantities sufficient to cover 100 square feet. This quantity is commonly termed "one square" by roofers and in trade literature. When ordering roofing material, it will be well to make allowance for waste such as in hips, valleys, and starter courses. This applies in general to all types of roofing.

39

The slope of the roof and the strength of the framing are the first determining factors in choosing a suitable covering. If the slope is slight, there will be danger of leaks with a wrong kind of covering, and excessive weight may cause sagging that is unsightly and adds to the difficulty of keeping the roof in repair. The cost of roofing depends to a great extent on the type of roof to be covered. A roof having ridges, valleys, dormers, or chimneys will cost considerably more to cover than one having a plain surface. Very steep roofs are also more expensive than those with a flatter slope, but most roofing materials last longer on steep grades than on low-pitched roofs. Frequently, nearness to supply centers permits the use, at lower cost, of the more durable materials instead of the commonly lower-priced, shorter-lived ones.

In considering cost, one should keep in mind maintenance and repair and the length of service expected from the building. A permanent structure warrants a good roof, even though the first cost is somewhat high. When the cost of applying the covering is high in comparison with the cost of the material, or when access to the roof is hazardous, the use of long-lived material is warranted. Unless insulation is required, semipermanent buildings and sheds are often covered with low-grade roofing.

Frequently, the importance of fire resistance is not recognized and sometimes it is wrongly stressed. It is essential to have a covering that will not readily ignite from glowing embers. Unless proper precautions against inside fires have been provided, a noncombustible roof is unnecessary except where it is exposed to sparks. The building regulations of many cities prohibit the use of certain types of roofings in congested areas where fires may spread rapidly. The National Board of Fire Underwriters has grouped many of the different kinds and brands of roofing in classes according to the protection afforded against spread of fire.

The appearance of a building can be changed materially by using the various coverings in different ways. Wood shingles and slate are often used to produce architectural effects. The roofs of buildings in a farm group should harmonize in color even though similarity in contour is not always feasible.

The action of the atmosphere in localities where the air is polluted with fumes from industrial works, or is saturated with salt (as along the seacoast), shortens the life of roofing made from certain metals. Sheet aluminum is particularly vulnerable to acid fumes.

All coal-tar pitch roofs should be covered with slag or a mineral coating, because when fully exposed to the sun, they deteriorate. Observation has shown that, in general, roofings with light-colored surfaces absorb less heat than those with dark surfaces. Considerable attention should be given to the comfort derived from a properly insulated roof. A thin uninsulated roof gives the interior little protection from heat in summer and cold in winter. Discomfort from summer heat can be lessened to some extent by ventilating the space under the roof. None of the usual roof coverings have any appreciable insulating value. If it is necessary to re-roof, consideration should be given to the feasibility of installing extra insulation under the roofing.

DETECTION OF ROOF LEAKS

A well constructed roof should be properly maintained. Periodic inspections should be made to detect breaks, missing shingles, choked gutters, damaged flashings, and also defective mortar joints of chimneys, parapets, coping, and such. At the first appearance of damp spots on the ceilings or walls, a careful examination of the roof should be made to determine the cause, and the defect should be promptly repaired. When repairs are delayed, small defects extend rapidly and involve not only the roof covering, but also the sheathing, framing, and interior.

Many of these defects can be readily repaired to keep water from the interior and to extend the life of the roof. Large defects or failures should be repaired by men familiar with the work. On many types of roofs, an inexperienced man can do more damage than he can do good. Leaks are sometimes difficult to find, but an examination of the wet spots on a ceiling furnishes a clue to the probable location. In some cases, the actual leak may be some distance up the slope. If near a chimney or exterior wall, the leaks are probably caused by a defective or narrow flashing, loose mortar joints, or dislodged coping. On flat roofs, the trouble may be the result of choked downspouts or an accumulation of water or snow on the roof higher than the flashing. Defective and loose flashing is not uncommon around scuttles, cupolas, and plumbing vent pipes. Roofing deteriorates more rapidly on a south exposure than on a north exposure, which is especially noticeable when wood or composition shingles are used.

Wet spots under plain roof areas are generally caused by holes in the covering. Frequently, the drip may occur much lower down the slope than the hole. Where attics are unsealed and roofing strips have been used, holes can be detected from the inside by light shining through. If a piece of wire is stuck through the hole it can be located from the outside.

Sometimes gutters are so arranged that when choked, they overflow into the house, or ice accumulating on the eaves will form a ridge that backs up melting snow under the shingles. This is a common trouble if roofs are flat and the eaves wide. Leaky downspouts permit water to splash against the wall and the wind-driven water may find its way through a defect into the interior. The exact method to use in repairing depends on the kind of roofing and the nature and extent of the defect.

SUMMARY

The roof of a building includes the roof cover which is protection against rain, snow, and wind, the sheathing which is a base for the roof cover, and the rafters which are the support for the entire roof structure.

The term "roofing" refers to the uppermost part of the roof. There are various types of roofing used, such as wood, which generally is in the form of shingles or shakes, metal or aluminum corrugated sheets, tile, roll roofing, asphalt-felt, synthetic plastics, and canvas.

Roll roofing is one of the easiest and most economical covering to use on low pitch roofs. This type of roofing can be applied directly to the wood sheathing, or it can be installed over roofing felt. When applying the felt, start the first strip at the eaves, allowing it to project ½ inch. The second, third, fourth, etc., strips, each should overlap the previous strip about 2 inches.

Various types and styles of flashing are used when a roof connects to any vertical wall, such as chimneys, outside walls, etc. Flashing around chimneys and skylights is installed in the same general manner as for vertical walls. It is generally made from roll roofing material or sheet metal bent to fit the contour of the vertical wall.

Canvas is often used for waterproofing boat decks and for sun decks subjected to foot traffic. Sheathing for this type of material must be smooth, with no splits or wide cracks. Canvas will quickly wear through at a turned-up edge or crack in the sheathing. With proper installation and by keeping it well painted, a canvas roof should last 25 to 30 years.

Sheet metal roofs are generally used on utility type buildings. The sheets are generally galvanized iron, tin, aluminum, copper, zinc, or monel metal. These types of roofing materials are comparatively light in weight and, if properly laid, will provide a watertight roof. Many times a corrugated sheet roof can be installed without the use of sheathing.

Most roofs require rain gutters and downspouts in order to carry the water to the sewer or outlet. Gutters and downspouts are usually built of metal, although wood gutters are used in some areas. Downspouts should be large enough to remove the water from the gutters. Much gutter deterioration is due to freezing water in low areas, rust, and restricted sections due to leaves or other debris.

REVIEW QUESTIONS

1. Name various types of roofing used.
2. What is flashing and why is it used?
3. Why can corrugated metal roofs be installed without roof sheathing?
4. How much coverage in square feet is "one square" of roofing?
5. What are some advantages in a canvas roof?

CHAPTER 2

Cornice Construction

The cornice is that projection of the roof at the eaves that forms a connection between the roof and the side walls. The three general types of cornice construction is the *box,* the *closed,* and the *open.*

BOX CORNICES

The typical box cornice shown in Fig. 1 utilizes the rafter projection for nailing surfaces for the facia and soffit boards. The soffit provides a desirable area for inlet ventilators. A frieze board is often used at the wall to receive the siding. In climates where snow and ice dams may occur on overhanging eaves, the soffit of the cornice may be sloped outward and left open 1/4 inch at the facia board for drainage.

CLOSED CORNICES

The closed cornice shown in Fig. 2 has no rafter projection. The overhang consists only of a frieze board and a shingle or crown moulding. This type is not so desirable as a cornice with a projection, because it gives less protection to the side walls.

WIDE BOX CORNICES

The wide box cornice in Fig. 3 requires forming members called lookouts, which serve as nailing surfaces and supports for the soffit board. The lookouts are nailed at the rafter ends and are also toenailed to the wall sheathing and directly to the stud. The soffit can be of various materials, such as beaded ceiling, plywood, or bevel siding. A bed moulding may be used at the juncture of the

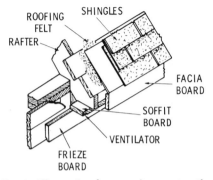

Fig. 1. Illustrating box cornice construction.

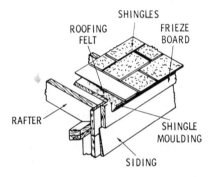

Fig. 2. Illustrating closed cornice construction.

soffit and frieze. This type of cornice is often used in hip roof houses, and the facia board usually carries around the entire perimeter of the house.

OPEN CORNICES

The open cornice shown in Fig. 4, may consist of a facia board nailed to the rafter ends. The frieze is either notched or cut out to fit between the rafters and is then nailed to the wall. The open cornice is often used for a garage. When it is used on a house, the roof boards are visible from below from the rafter ends to the wall line, and should consist of finished material. Dressed or matched V-beaded boards are often used.

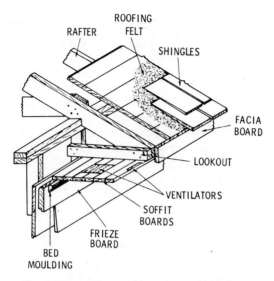

Fig. 3. Illustrating wide cornice construction.

CORNICE RETURNS

The cornice return is the end finish of the cornice on a gable roof. The design of the cornice return depends to a large degree on the rake or gable projection, and on the type of cornice used.

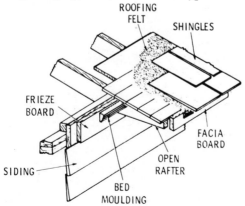

Fig. 4. Illustrating open cornice construction.

In a close rake (a gable end with very little protection), it is necessary to use a frieze or rake board as a finish for siding ends, as shown in Fig. 5. This board is usually 1-1/8 inches thick and follows the roof slope to meet the return of the cornice facia. Crown moulding or other type of finish is used at the edge of the shingles.

When the gable end and the cornice have some projection as shown in Fig. 6, a box return may be used. Trim on the rake projection is finished at the cornice return. A wide cornice with a small gable projection may be finished as shown in Fig. 7. Many variations of this trim detail are possible. For example, the frieze board at the gable end might be carried to the rake line and mitered with a facia board of the cornice. This siding is then carried across the cornice end to form a return.

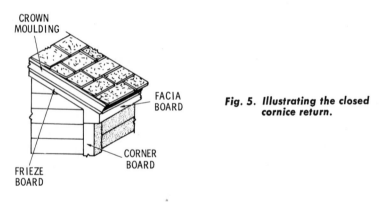

CROWN MOULDING

FACIA BOARD

CORNER BOARD

FRIEZE BOARD

Fig. 5. Illustrating the closed cornice return.

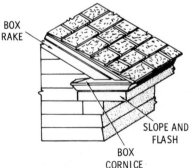

BOX RAKE

SLOPE AND FLASH

BOX CORNICE

Fig. 6. Illustrating the box cornice return.

48

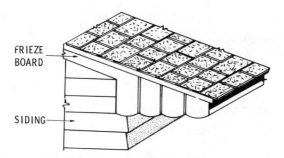

FRIEZE BOARD

SIDING

Fig. 7. Illustrating the wide cornice return.

RAKE OR GABLE-END FINISH

The rake section is that trim used along the gable end of a house. There are three general types commonly used; the *closed,* the *box with a projection,* and the *open.* The closed rake, as shown in Fig. 8, often consists of a frieze or rake board with a

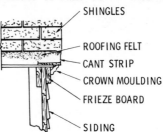

SHINGLES

ROOFING FELT

CANT STRIP

CROWN MOULDING

FRIEZE BOARD

SIDING

Fig. 8. Illustrating the closed end finish.

crown moulding as the finish. A 1 × 2 inch square edge moulding is sometimes used instead of the crown moulding. When fiber board sheathing is used, it is necessary to use a narrow frieze board that will leave a surface for nailing the siding into the end rafters.

If a wide frieze is used, nailing blocks must be provided between the studs. Wood sheathing does not require nailing blocks. The trim used for a box rake section requires the support of the projected roof boards, as shown in Fig. 9. In addition, lookouts or nailing biocks are fastened to the side wall and to the roof sheathing. These lookouts serve as a nailing surface for both the soffit and the facia boards. The ends of the roof boards are nailed to the facia. The frieze board is nailed to the side wall studs, and

49

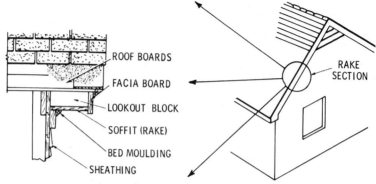

Fig. 9. Illustrating the box end finish.

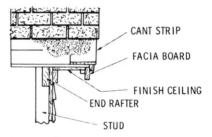

Fig. 10. Illustrating the open end finish.

and the crown and bed mouldings complete the trim. The underside of the roof sheathing of the open projected rake as shown in Fig. 10, is generally covered with liner boards such as 5/8 inch beaded ceiling. The facia is held in place by nails through the roof sheathing.

SUMMARY

The cornice is that part of the roof at the eaves that forms a connection between the roof and side walls. There are generally three styles of cornice construction called box, closed, and open.

The box cornice construction generally uses the rafter ends as a nailing surface for the facia and soffit board. A board called the "frieze board" is used at the wall to start the wood siding. Wide box cornices require framework called lookouts, which serves as nailing surfaces and support the soffit board. The lookouts are nailed at the rafter end and also nailed at the other end to the wall stud.

On the closed cornice, there is no rafter projection. There is no overhang, using only a frieze board and moulding. There is no protection from the weather for the side walls with this type of construction.

REVIEW QUESTIONS

1. Name the three types of cornice construction.
2. What is a frieze board?
3. Explain the purpose of the facia board.
4. What is the lookout block and when is it used?
5. What is the soffit board?

CHAPTER 3

Miter Work

In treating this important branch of carpentery, it is advisable to mention some of its elements. By definition, a miter is *the joint formed by two pieces of moulding, each cut at an angle so as to match when joined angularly;* also, to miter means *to meet and match together on a line bisecting the angle of junction, especially at a right angle,* in other words, *to cut and join together the ends of two pieces obliquely at an angle.*

MITER TOOLS

To do this with precision, the proper tools are necessary. The first is, of course the saw, which should be a good 20-inch back saw of about eleven or twelve teeth to the inch, filed to a keen edge and rubbed off on the sides with the face of an oil stone. For precision, a manufactured metal miter box should be used. However, a serviceable miter box, such as shown in Fig. 1, can be made of suitable hardwood by the carpenter for most of the common miter cuts.

Although several patent miter boxes are now in use, the wooden box will always have its place where the most common miters are to be cut. The steel miter box is recommended where precision work is needed. Some of the metal miter boxes have attachments whereby frames may be held firmly in position while the miter is nailed. There are also hand and foot power miter cutters by which both miters are rapidly cut in one operation.

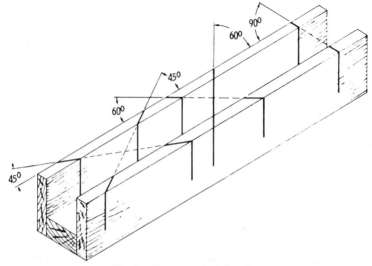

Fig. 1. Home-made miter box.

MOULDINGS

In the ornamental side of carpentry construction, various forms of fancy-shaped strips, called mouldings, are used. Some of these are designed to lay flush or flat against the surfaces to which they are attached, as in Fig. 3; others are shaped to lie inclined at an angle to the nailing surfaces, as in Fig. 4. It is the *rake* or *spring* type moulding that is hard to cut.

Mitering Flush Mouldings

Where two pieces of moulding join at right angles, as for instance the sides of a picture frame, the miter angle is 45°. The term "miter angle" means the angle formed by the miter cut and edge of the moulding, as in Fig. 5.

In paneling for a stairway, the mouldings are joined at various angles, as in Fig. 6. This is known as varying miters, and a problem arises to find the miter angles. This is easily done by remembering that the miter angle is always half of the joint angle. To find the miter cut, that is the angle at which the miter cut is made, bisect the joint angle. This is done as explained in Fig. 7. The triangle *ABC* corresponds to the triangle *A* in Fig. 6. To find

54

the miter cut at *A*, describe the arc *MS* of any radius with *A* as center. With *M* and *S* as centers, describe arcs *L* and *F*, intersecting at *R*. Draw line *AR*, which is the miter cut required.

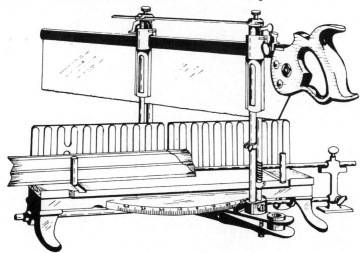

Fig. 2. A typical metal miter box with graduated scales of angle.

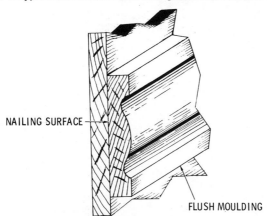

Fig. 3. Flush type moulding.

Mitering Spring Mouldings

A spring moulding is one that is made of thin material, and is leaned or inclined away from the nailing surface, as explained in

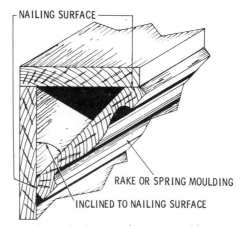

Fig. 4. Spring or rake type moulding.

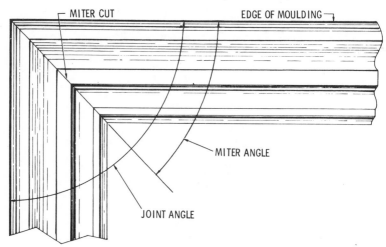

Fig. 5. Two pieces of flush moulding joined at 90°.

Fig. 4. These mouldings are difficult to miter, especially when the joint is made with a gable, springs, or raking moulding. The two most unusual forms of miters to cut on spring mouldings are those on the inside and outside angles as shown in Fig. 8. The pieces are represented as they would appear from the top sill looking down.

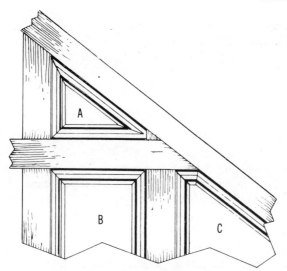

Fig. 6. Panel work of a wall illustrating varying miters. In panel (A) each angle is different, (B) both miters are equal, (C) has two different angles.

A difficult operation for most carpenters is the cutting of a spring moulding when the horizontal portion has to miter with a gable or raking moulding. The miter-box cuts for such joints are laid out as shown in Fig. 9. To lay out these cuts in constructing the miter box, make the "down cuts" *BB*, the same pitch as the plumb cut on the rake. The "over cuts" *OO* and *O'O'*, should be

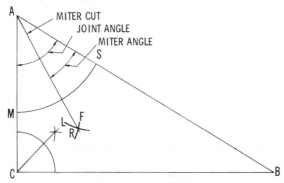

Fig. 7. A method of finding various miter cuts for different angles.

obtained as follows. Suppose a roof has a quarter pitch, find the rafter inclination as in Fig. 10, by laying off $AB = 12$ in. run and $BC = 6$ in. rise, giving the roof angle CAB for 1/4 pitch and rafter length $AC = 13.42$ in. per foot run. With the setting 13.42 and 12, lay the steel square on top of the miter box, as shown in Fig. 11.

Mitering Panel and Raised Mouldings

The following instructions illustrate how raised and rabbetted mouldings may be cut and inserted in panels. Fig. 12 shows a panel and moulding designed for a room or wardrobe door. AB denotes the outside frame, and C, the raised panel. D and E is the pine fillets inserted in the plowing, and F is the panel moulding which has to be mitered around the inside edges of the frame. Point G is the rabbet or lips on the moulding F. If the framing AB is carefully constructed, and the surfaces are equal, the offset down to the panel will be equal all around, then all that is necessary is to make a hardwood strip or saddle equal in width to the depth of the offset.

The front door shown in Fig. 13 has both flush and raised panels. A raised 1-inch moulding is on the outside or street side and an

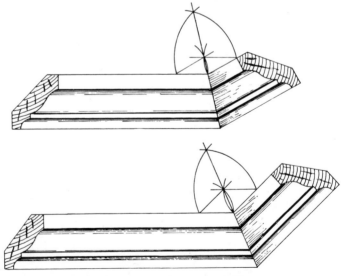

Fig. 8. Illustration of the two most unusual miter forms to cut.

ordinary ogee and chamfer is on the inside. The enlarged section is shown. This door is a good example of mitered mouldings which forms an attractive design. The difference between outside and inside miters must be explained—an *inside* miter is one in which the profile of the moulding is contained, or rather the outside lines and highest parts are contained, within the angle of the framing. An *outside* miter is one which is directly opposite and not contained, but the whole of the moulding is mitered on the panel outside the angle. Both miters are sawed similarly in the box with the exception of the reversing of the intersections.

Cutting Long Miters

In numerous instances, miter cuts must be made that cannot be cut in an ordinary or patent miter box. In such cases the work is

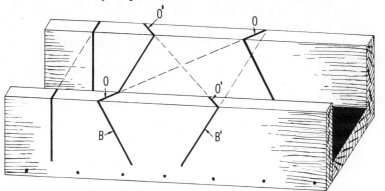

Fig. 9. Miter box lay-out for cutting a spring moulding when the horizontal portion has to miter with a gable or raking moulding.

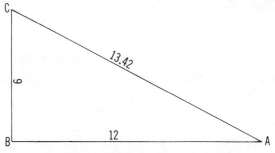

Fig. 10. Method of finding the angle for the cuts shown in Fig. 9.

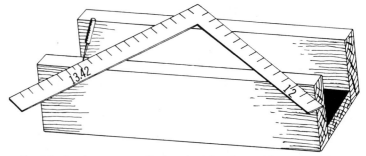

Fig. 11. Steel square applied to the miter box with 13.42 and 12 setting to mark for cuttings.

facilitated by making a special box if there is several cuts of a kind to be made.

Fig. 14 shows a box 13 inches high and has a flare of 3-1/4 inches. Its construction requires miter cuts which cannot be made on an ordinary miter box. One corner is a rabbet joint and the

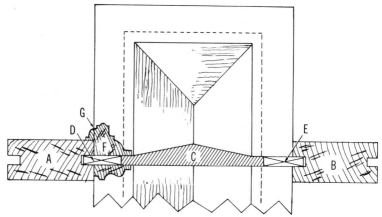

Fig. 12. Illustrating a panel and moulding design.

other corner is a miter joint. Each corner can be cut out by the use of an adjustable-table power saw.

Coping

By definition cope means to cover, or match against, a covering. Coping is generally used for mouldings, the square and flat surfaces

Fig. 13. End and side view of a door with raised moulding.

being fitted together, one piece abutting against the other. Against plaster, the inside miter is useless since one piece is almost certain to draw away and open the joint as it is being nailed into the studding. It can be mitered tight enough by cutting the lengths a little full and springing them into place, but it is not advisable except possibly in solid corners. If against plastered walls, it may crack. The best way to make this joint is to cope it.

Fig. 15 shows various types of coped joints. In order to obtain this joint, the piece of moulding is placed in a miter box and cut to a 45° angle. After this is done, the miter angle is cut by a coping saw along the design of the moulding. If the corners are square, the miter and coped joints will fit perfectly.

Fig. 16 shows that when a moulding is cut in a miter box for coping, it is always the reverse of the profile, and when cut out to

61

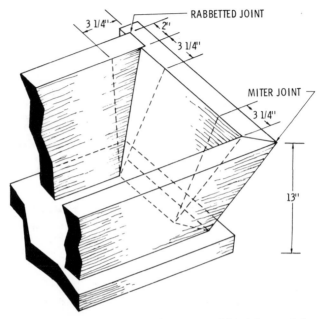

Fig. 14. A view of two joints, one showing a rabbet joint, and the other a miter joint.

the line thus formed, preferably with a coping saw, it fits to it at every inside corner so as to be invisible. In brief, each curved line and members join and intersect each to each without interruption at any point.

SUMMARY

Various precision tools are necessary to make proper miter angles. One such tool is a good miter box with the proper type saw. A serviceable miter box can be made from suitable hardwood by a carpenter for most of the common miter cuts.

In carpentry construction, various forms of fancy-shaped mouldings are used. Some are designed to lay flat or flush against the surface and others are shaped to lie inclined or at an angle to the surface.

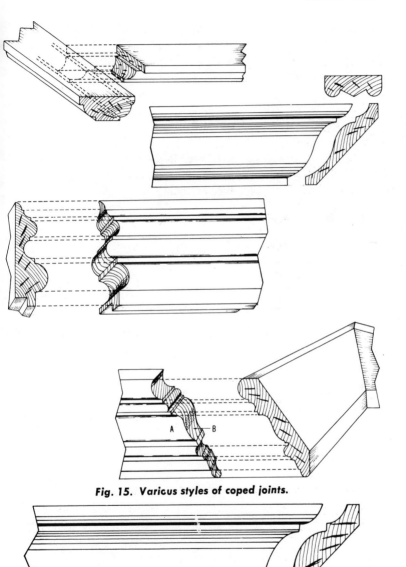

Fig. 15. Various styles of coped joints.

Fig. 16. Coped crown or spring moulding.

Coping is another form of joining angles. After the moulding is cut at a 45° angle, a coping saw is used to saw along the design of the adjoining moulding. This is done to relieve any possible strain in moulding corners on walls, such as plaster.

REVIEW QUESTIONS

1. What is a miter box?
2. What type of saw should be used in miter cuts?
3. What is coping?
4. What is spring or rake type moulding?
5. Explain the difference between a rabbet joint and a miter joint.

CHAPTER 4

Doors

Doors, both exterior and interior, are classified as four types—
batten, panel, flush, and louver. The batten door can be made in
several ways, one of the simplest consisting of diagonal boards
nailed together as two layers, each layer at right angles to the other
as shown in Fig. 1. Constructed in this manner, this type of door
frequently is used as the core for metal-sheathed fire doors.

Another type of batten door is made up of vertical tongue-
and-groove or shiplap boards held rigid by crosspieces. These
crosspieces are from two to four in number, and may or may not
be diagonally braced. The crosspieces are called ledgers. The
ledgers are placed with their edge 6 inches from the ends of the
door boards. A diagonal board is placed between the ledgers,
beginning at the top ledger end opposite the hinge side of the door
and running to the lower ledger diagonally across the door, as
shown in Fig. 2.

MANUFACTURED OR MILL-MADE DOORS

For all practical purposes, doors can be obtained from the mill
in stock sizes much cheaper than they can be made by hand.
Stock sizes of doors cover a wide range, but those most commonly
used are $2'4'' \times 6'8''$, $2'8'' \times 6'8''$, $3'0'' \times 6'8''$, and $3'0'' \times 7'0''$. These sizes are either 1-3/8 or 1-3/4 inches thick.

Paneled Doors

Paneled doors are made in a variety of panel arrangements,
both horizontal, vertical, and combinations of both. A single-panel

door has for its component parts a top rail, bottom rail, and two stiles which form the sides of the door. Panels of the horizontal type have intermediate rails forming the panels; and panels of the vertical type have horizontal rails and vertical stiles forming the panels.

The rails and stiles of a door are generally mortised and tenoned, the mortise being cut in the side stiles as shown in Fig. 3. Top and bottom rails on paneled doors differ in width, the bottom rail being considerably wider. Intermediate rails are usually the same width as the top rail. Paneling material is usually plywood which is set in grooves or dadoes in the stiles and rails, with the moulding attached on most doors as a finish.

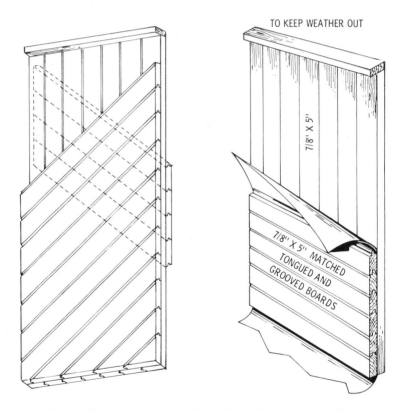

Fig. 1. Illustrating two- and three-ply hand-made batten doors.

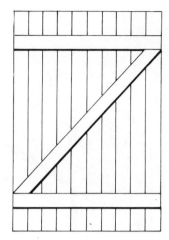

Fig. 2. Another type of batten door using a ledger board.

Flush Doors

Flush doors are usually perfectly flat on both sides. Solid planks are rarely used for flush doors. Flush doors are made up with solid or hollow cores with two or more plies of veneer glued to the cores.

Solid-Core Doors

Solid-core doors are made of short pieces of wood glued together with the ends staggered very much like in brick laying. One or two piles of veneer are glued to the core. The first section, about 1/8 inch thick, is applied at right angles to the direction of the core, and the other section, 1/8 inch or less, is glued with the grain vertical.

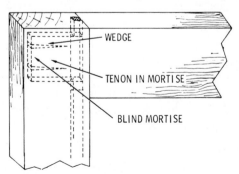

Fig. 3. Door construction showing mortise joints.

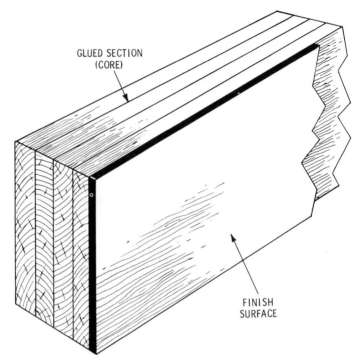

GLUED SECTION
(CORE)

FINISH
SURFACE

Fig. 4. Construction of a laminated or veneered door.

A 3/4-inch strip, the thickness of the door is glued to the edges of the door on all four sides. This type of door construction is shown in Fig. 4.

Hollow-Core Doors

Hollow-core doors have wooden grids or other honeycomb material for the base, with solid wood edging strips on all four sides. The face of this type door is usually 3-ply veneer instead of two single plies. The hollow-core door has a solid block on both sides for installing door knobs and to permit the mortising of locks. The honeycomb-core door is for *interior* use only.

Louver Doors

This type of door has either stationary or adjustable louvers, and may be used as an interior door, room divider, or a closet

Fig. 5. Various styles of louver doors.

door. The louver door comes in many styles, such as shown in Fig. 5. An exterior louver door may be used, which is called a *jalousie* door. This door has the adjustable louvers usually made of wood or glass. Although there is little protection against winter winds, a solid-type storm window is made to fit over the louvers to give added protection.

INSTALLING MILL-BUILT DOORS

There are numerous ways in which a door frame may be constructed. A door frame consists of the following essential parts.

1. Sill.
2. Threshold.
3. Side and top jamb.
4. Casing.

These essential parts are shown in Fig. 6.

Door Frames

Before the exterior covering is placed on the outside walls, the door openings are prepared for the frames. To prepare the openings, square off any uneven pieces of sheathing and wrap heavy building paper around the sides and top. Since the sill must be worked into a portion of the subflooring, no paper is put on the floor. Position the paper from a point even with the inside portion of the stud to a point about 6 inches on the sheathed walls and tack it down with small nails.

Outside door frames are constructed in several ways. In more hasty constructions, there will be no door frame. The studs on each side of the opening acts as the frame and the outside casing is applied to the walls before the door is hung. The inside door frame is constructed in the same manner as the outside frame.

Door Jambs

Door jambs are the lining to the framing of a door opening. Casings and stops are nailed to the jamb, and the door is securely fastened by hinges at one side. The width of the jamb will vary in accordance with the thickness of the walls. The door jambs are made and set in the following manner.

1. Regardless of how carefully the rough openings are made, be sure to plumb the jambs and level the heads when the jambs are set.
2. Rough openings are usually made 2-1/2 inches larger each way than the size of the door to be hung. For example, a 2'8" × 6'8" door would need a rough opening of 2'10-1/2" × 6'10-1/2". This extra space allows for the jamb, the wedging, and the clearance space for the door to swing.
3. Level the floor across the opening to determine any variation in floor heights at the point where the jamb rests on the floor.
4. Cut the head jamb with both ends square, allowing for the width of the door plus the depth of both dadoes and a full 3/16 inch for door clearance.
5. From the lower edge of the dado, measure a distance equal to the height of the door plus the clearance wanted at the bottom.

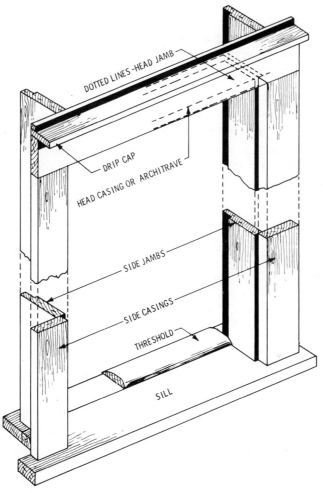

Fig. 6. View of a door frame showing the general construction.

6. Do the same thing on the opposite jamb, only make additions or subtractions for the variation in the floor.
7. Nail the jambs and jamb heads together through the dado into the head jamb, as shown in Fig. 7.
8. Set the jambs into the opening and place small blocks under each jamb on the subfloor just as thick as the finish floor

71

will be. This will allow the finish floor to go under the door.

9. Plumb the jambs and level the jamb head.
10. Wedge the sides to the plumb line with shingles between the jambs and the studs, and then nail securely in place.
11. Take care not to wedge the jambs unevenly.
12. Use a straightedge 5 to 6 feet long inside the jambs to help prevent uneven wedging.
13. Check each jamb and the head carefully. If a jamb is not plumb, it will have a tendency to swing the door open or shut, depending on the direction in which the jamb is out of plumb.

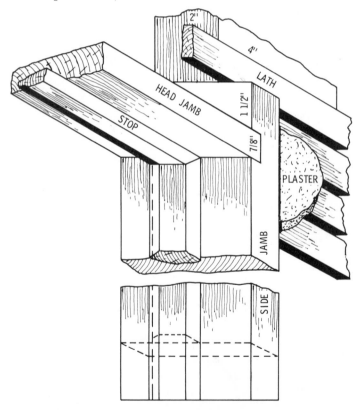

Fig. 7. Details showing the upper head jamb dadoed into the side jamb.

Door Trim

Door trim material is nailed onto the jambs to provide a finish between the jambs and the plastered wall. This is frequently called *casing*. Sizes vary from 1/2 to 3/4 inches in thickness, and from 2-1/2 to 6 inches in width. Most casing material has a concave back, to fit over uneven plaster. In miter work, care must be taken to make all joints clean, square, neat, and well fitted. If the trim is to be mitered at the top corners, a miter box, miter square, hammer, nail set, and block plane will be needed. Door openings are cased up in the following manner.

1. Leave a 1/4-inch margin between the edge of the jamb and the casing on all sides.
2. Cut one of the side casings square and even with the bottom of the jamb.
3. Cut the top or mitered end next, allowing 1/4 inch extra length for the margin at the top.
4. Nail the casing onto the jamb and set it even with the 1/4-inch margin line, starting at the top and working toward the bottom.
5. The nails along the outer edge will need to be long enough to penetrate the casing, plaster, and wall stud.
6. Set all nail heads about 1/8 inch below the surface of the wood.
7. Apply the casing for the other side of the door opening in the same manner, followed by the head (or top) casing.

HANGING MILL-BUILT DOORS

If mill-built doors are used, install them in the finished door opening as described below.

1. Cut off the stile extension, if any, and place the door in the frame. Plane the edges of the stiles until the door fits tightly against the hinge side and clears the lock side of the jamb about 1/16 inch. Be sure that the top of the door fits squarely into the rabbeted recess and that the bottom swings free of the finished floor by about 1/2 inch. The

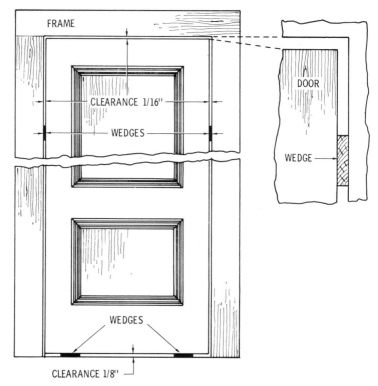

Fig. 8. Sizing a door for an opening.

lock stile of the door must be beveled slightly so that the edge of the door will not strike the edge of the door jamb.

2. After the proper clearance of the door has been made, set the door in position and place wedges as shown in Fig. 8. Mark the position of the hinges on the stile and on the jamb with a sharp pointed knife. The lower hinge must be placed slightly above the lower rail of the door. The upper hinge of the door must be placed slightly below the top rail in order to avoid cutting out a portion of the tenons of the door rails. There are three measurements to mark— the location of the butt hinge on the jamb, the location of the hinge on the door, and the thickness of the hinge on both the jamb and the door.

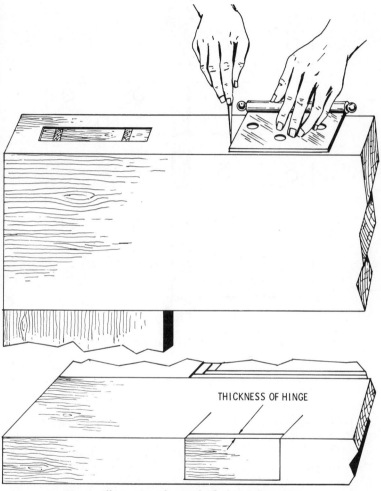

Fig. 9. Illustrating the method of installing hinges.

3. Door butt hinges are designed to be mortised into the door and frame, as shown in Fig. 9. Fig. 10 shows a new type of hinge which is installed directly to the door and jamb. Three hinges are usually used on full-length doors to prevent warping and sagging.

4. Using the butt as a pattern, mark the dimension of the

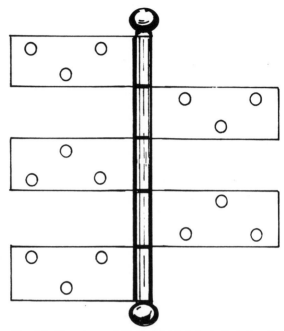

Fig. 10. New type of hinge which is installed directly to the door and jamb.

butts on the door edge and the face of the jamb. The butts must fit snugly and exactly flush with the edge of the door and the face of the jamb.

After placing the hinges and hanging the door, mark off the position for the lock and handle. The lock is generally placed about 36 inches from the floor level. Hold the lock in position on the stile and mark off with a sharp knife the area to be removed from the edge of the stile. Mark off the position of the door-knob hub. Bore out the wood to house the lock and chisel the mortises clean. After the lock assembly has been installed, close the door and mark the jamb for the striker plate.

SWINGING DOORS

Frequently, it is desirable to hang a door so that it opens as you pass through from either direction, yet remains closed at all other times. For this purpose, swivel-style spring hinges have been perfected. This type of hinge attaches to the rail of the door and to the jamb like an ordinary butt hinge. Another type is mortised into the bottom rail of the door and is fastened to the floor with a floor plate. In most cases, the floor-plate hinge, as shown in Fig. 11, is best, because it will not weaken and let the door sag. It is also designed with a stop to hold the door open at right angles, if so desired.

SLIDING DOORS

Sliding doors are usually used for walk-in closets. They take up very little space, and they also allow a wide variation in floor plans.

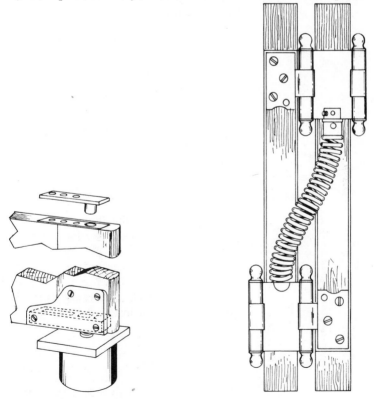

Fig. 11. Two kinds of swivel-type spring hinges.

This type of door usually limits the access to a room or closet unless the doors are pushed back into a wall. Very few sliding doors are pushed back into the wall because of the space and expense involved. Fig. 12 shows a double and a single sliding door track.

GARAGE DOORS

Overhead garage doors are made in a variety of sizes and designs. The principal advantage in using the overhead-type garage door, is the disappearing qualities. The door can be opened and rolled up out of the way, affording maximum convenience and efficiency. In addition, the door cannot be blown shut due to wind, and it is not obstructed by snow and ice.

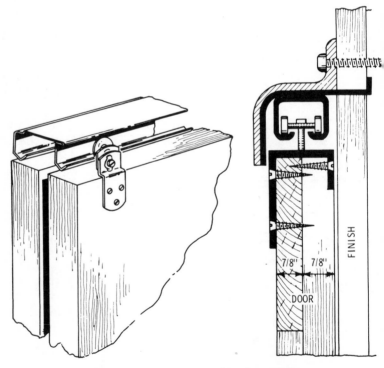

Fig. 12. Two types of sliding-door tracks.

Although designed primarily for use in residential and commercial garages, doors of this type are also employed in service stations, factory receiving docks, boathouses, and many other buildings. In order to permit overhead-door operation, garage doors of this type are built in suitable hinged sections. Usually 4 to 7 sections are used, depending upon the door height requirements. Standard residential garage doors are usually 9′ × 7′ for singles

Courtesy of Overhead Door Corporation

Fig. 13. Typical 18-foot overhead residential garage door.

and 16′ × 7′ for a double. Residential-type garage doors are usually manufacture 1-3/4 inches thick unless otherwise requested.

When ordering doors for the garage, the following information should be forwarded to the manufacturer:

1. Width of opening between the finished jambs.
2. Height of the ceiling from the finished floor under the door to the underside of the finished header.
3. Thickness of the door.
4. Design of the door (number of glass windows and sections).
5. Material of jambs (they must be flush).
6. Head room from the underside of the header to the ceiling, or to any pipes, lights, etc.

79

7. Distance between the sill and the floor level.
8. Proposed method of anchoring the horizontal track.
9. Depth to the rear from inside of the upper jamb.
10. Inside face width of the jamb buck, angle, or channel.

This information applies for overhead doors only, and does not apply to garage doors of the slide, folding, or hinged type. Doors can be furnished to match any style of architecture and may be provided with suitable size glass windows if desired (Fig. 13).

If your garage is attached to your house, your garage door often represents from one-third to one-fourth of the face of your house.

(A) Fiberglass.   (B) Steel.

(C) Wood.

Fig. 14. Illustrating three types of garage doors.

Style and material should be considered to accomplish a pleasant effect with masonry or wood architecture. Fig. 14 shows three types of overhead garage doors that can be used with virtually any kind of architectural design. Many variations can be created from com-

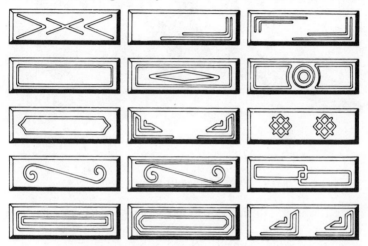

Fig. 15. Variations in carved or routed panel designs.

binations of raised panels with routed or carved designs as shown in Fig. 15. These panels may also be combined with plain raised panels to provide other dramatic patterns and color combinations.

Automatic garage-door openers in the past have been a luxury item, but in the past few years the price has been reduced and failure minimized to the extent that most new installations include this feature. Automatic garage-door openers save time, eliminates the need to stop the car and get out in all kinds of weather, and you also save the energy and effort required to open and close the door by hand.

The automatic door opener is a radio-activated motor-driven power unit that mounts on the ceiling of the garage, and attaches to the inside top of the garage door. Electric impulses from a wall-mounted push button, or radio waves from a portable radio transmitter in your car, starts the door mechanism. When the door reaches its limit of travel (up or down), the unit turns itself off and awaits your next command. Most openers on the market have a safety factor built in. If the door encounters an obstruction

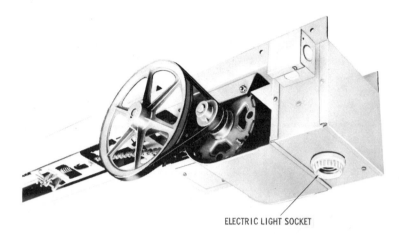

ELECTRIC LIGHT SOCKET

Courtesy of Overhead Door Corporation
Fig. 16. Typical automatic garage-door opener.

in its travel, it will instantly stop, or stop and reverse its travel. The door will not close until the obstruction has been removed. When the door is completely closed, it is automatically locked and cannot be opened from the outside, making it burglarproof. Fig. 16 shows a typical automatic garage-door opener which can be quickly disconnected for manual-door operation in case of electrical power failure. Notice the electric light socket on the automatic opener unit, which turns on when the door opens to light up the inside of the garage.

SUMMARY

Most doors, both exterior and interior, are classified as four types—batten, panel, flush, and louver. This type of door is very popular as a core for metal-sheathed fire doors.

Paneled doors are made in many styles. The rails and stiles are generally mortised and tenoned. Top and bottom rails on paneled doors differ in width, with the bottom rail considerably wider. The center rail is generally the same width as the top rail. The panel material is usually plywood which is set in grooves or dadoes in the stiles and rails.

Flush doors, hollow and solid-core, generally all have the same appearance. The main difference in these is the method of construction. Solid-core doors are made of short pieces of wood glued together with the ends staggered very much like brick laying. Hollow-core doors have wooden grids or some type of honeycomb material for the base, with solid wood edging strips on all four sides. Glued to the cores of these doors are two or three layers of wood veneer which make up the door panel. The honeycomb-core door is made for interior use only.

REVIEW QUESTIONS

1. Name the various type doors.
2. Why are honeycomb-core doors made for interior use only?
3. What is a door stop?
4. When hanging a door, how much clearance should there be at top, bottom, and sides?
5. How are solid-core doors constructed?

CHAPTER 5

Windows

Windows are generally classified as sliding, double-hung, and casement. All windows, whatever the type, consist essentially of two parts, the frame and the sash. The frame is made up of four basic parts—the head, two jambs, and the sill. Good construction around the window frame is essential to good building. Where openings are to be provided, studding must be cut away and its equivalent strength replaced by doubling the studs on each side of the opening to form trimmers, and inserting a header at the top. If the opening is wide, the header should be doubled and trussed. At the bottom of the opening, a header or rough sill is inserted.

WINDOW FRAMING

This is the frame into which the window sash fits. It is set into a rough opening in the wall framing, and is intended to hold the sash in place.

DOUBLE-HUNG WINDOWS

The double-hung window is made up of two parts—an upper and lower sash which slide vertically past each other. An illustration of this type of window is shown in Fig. 2. This type of window has some advantages and some disadvantages. Screens can be installed on the outside of the window without interfering with its operation. For full ventilation of a room, only one half of the area

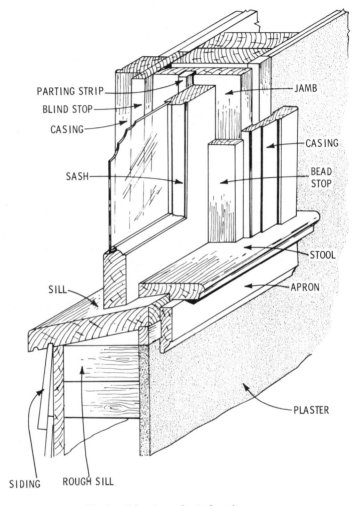

Fig. 1. Side view of window frame.

of the window can be utilized, and any current of air passing across its face is, to some extent, lost in the room. Double-hung windows are sometimes more involved in their frame construction and operation than the casement window. Ventilation fans and air conditioners can be placed in the window with it partly closed.

HINGED OR CASEMENT WINDOWS

There are basically two types of casement windows—the out-swinging and the inswinging. These windows may be hinged at the side, top, or bottom. The casement window which opens out requires the screen to be located on the inside. This type of window, when closed, is most efficient as far as waterproofing. The inswinging, like double-hung windows, are clear of screens, but they are extremely difficult to make watertight. Casement windows have the advantage of their entire area being opened to air currents, thus catching a parallel breeze and slanting it into a room. Casement windows are considerably less complicated in their construction than double-hung units. Sill construction is very much like that for a double-hung window, however, but with the stool much wider and forming a stop for the bottom rail. When there are two casement windows in a row in one frame, they are separated by a vertical double jamb called a mullion, or the stiles may come together in pairs like a french door. The edges of the stiles may be a reverse rabbet, a beveled reverse rabbet with battens, or beveled astrogals. The battens and astrogals insure better weathertightness. Fig. 3 shows a typical casement window with a mullion.

WINDOW SASH

Most windows are normally composed of an upper and a lower sash. These sash slide up and down, swing in or out, or may be stationary. There are two general types of wood sash—fixed or permanent, and movable. Fixed window sash are removable only with the aid of a carpenter. Movable sash may be of the variety that slides up and down in channels in the frame, called *double-hung*. Casement-type windows swing in or out and are hinged on the sides. Some sliding sash are counterbalanced by weights, called *sash weights,* their actual weight being equal to one-half that of each sash. Sash are classified according to the number of lights—single or double.

Sash Installation

Place the upper double-hung sash in position and trim off a slight portion of the top rail to insure a good fit, and tack the

87

Fig. 2. Illustration of a double-
hung window.

upper sash in position. Fit tne lower sash in position by trimming
off the sides. Place the lower sash in position, and trim off a
sufficient amount from the bottom rail to permit the meeting rails

Fig. 3. Casement type window.

to meet on a level. In most cases, the bottom rail will be trimmed on an angle to permit the rail and sill to match both inside and outside, as shown in Fig. 4.

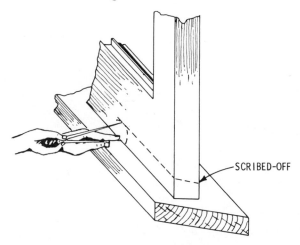

SCRIBED-OFF

Fig. 4. Illustrating bottom rail trim to match sill plate.

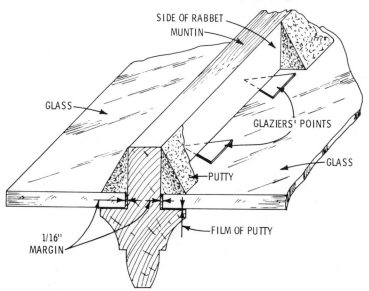

SIDE OF RABBET
MUNTIN
GLASS
GLAZIERS' POINTS
GLASS
PUTTY
FILM OF PUTTY
1/16"
MARGIN

Fig. 5. The glazier points which are removed to replace broken glass.

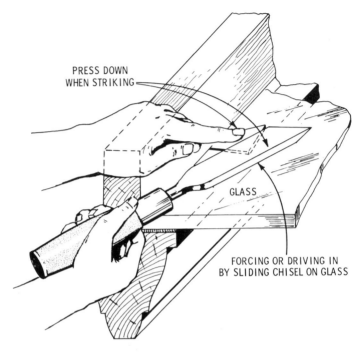

PRESS DOWN
WHEN STRIKING

GLASS

FORCING OR DRIVING IN
BY SLIDING CHISEL ON GLASS

Fig. 6. Illustrating the proper way to install glazier points.

Sash Weights

If sash weights are used, remove each sash after it has been properly cut and sized. Select sash weights equal to one-half the weight of each sash and place in position in the weight pockets. Measure the proper length of sash cord for the lower sash and attach it to the stiles and weights on both sides. Adjust the length of the cord so that the weight will not strike the pulley or bottom of the frame when window is moved up and down. Install the cords and weights for the upper sash and adjust the cord so that the weights run smoothly. Close the pockets in the frame and install the blind stop, parting strip, and bead stop.

There are many other types of window lifts, such as spring-loaded steel tapes, spring-tension metal guides, and full-length coil springs.

GLAZING SASH

The glass or *lights,* as they are called, are generally cut 1/8 inch smaller on all four sides to allow for irregularities in cutting and in the sash. This leaves an approximate margin of 1/16 inch between the edge of the glass and the sides of the rabbet. Fig. 5 shows two lights or panes of glass in position for glazing. To install the window glass properly, first spread a film of soft putty close to the edge on the inside portion of the glass. After the glass has been inserted, drive or press in at least two glazier points on each side. This is illustrated in Fig. 6.

After the glass is firmly secured with the glazier points, the putty (which is soft), is put on around the glass with a putty knife as shown in Fig. 7. Do not project the putty beyond the

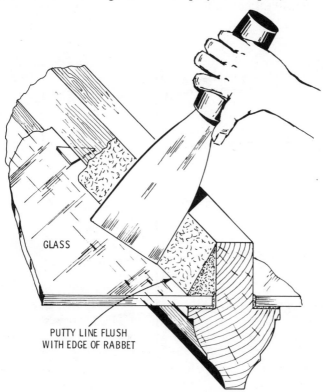

GLASS

PUTTY LINE FLUSH
WITH EDGE OF RABBET

Fig. 7. Illustrating the proper way to install window putty.

91

edge of the rabbet so that it will be visible from the other side.

Putty is usually purchased in a can with a lid which provides an air-tight seal. The putty should be soft and pliable to work properly with a putty knife. This can be accomplished by working the putty in your hands, or by mixing a small amount of linseed oil to the putty. After the putty has hardened, it should be painted to match the window sash.

WINDOW SCREENS

Window screen frames are usually 1-3/4 or 2-1/4 inches wide. The screen may be attached by stapling or tacking. The frame corners may be constructed with an open mortise, with the rails tenoned into the stiles, with half-lap corners, or with butt joints

Fig. 8. Showing the movable type shutters.

with corrugated fasteners. In either case, the joints may be nailed and glued. When attaching the screen material, start at one end and tack or staple it with copper staples, holding the material tight as you nail. Hand stretch the screen along the side, working toward the other end and attach, making sure that the weave is parallel to the ends and sides. Tack the sides and then apply the moulding. Copper staples should be used for bronze or copper screen; cadmium staples should be used for aluminum screens.

In most cases, factory built combination aluminum storm windows are installed. The combination storm windows are attractive and can be made to fit most size openings.

SHUTTERS

In coastal areas where damaging high winds occur frequently, shutters are necessary to protect large plate-glass windows from being broken. The shutters are mounted on hinges, and can be closed at a moments notice. Throughout the midwest, shutters are generally installed for decoration only, and are mounted stationary to the outside wall. There are generally two types of shutters—the solid panel, and the slat or louver type. Louver shutters can have stationary or movable slats, as shown in Fig. 8.

SUMMARY

Many styles and sizes of windows are used in various house designs, but windows are generally classified as sliding, double-hung, or casement. A window consists essentially of two parts, the frame and the sash.

Good construction around window frames is essential to good building. Where window openings are to be provided, studding must be cut and its equivalent strength replaced by doubling the studs on each side of the opening. The top and bottom of each opening must have a double header to add strength. The rough window opening is generally made at least 10 inches larger each way than the window glass size. This extra 10-inch allowance provides room for weights, springs, or balances.

Double-hung windows are made up of two parts—the upper and lower sash which slide vertically past each other. Only one-half of the area of the window can be used for ventilation, which is a disadvantage.

REVIEW QUESTIONS

1. Name the various window classifications.
2. What size should the rough opening be for a double-hung window?
3. What are some advantages of casement windows?
4. Name a few advantages in using window shutters.
5. What are glazier points, and why should they be used when installing window glass?

CHAPTER 6

Sheathing and Siding

Sheathing is nailed directly to the framework of the building. Its purpose is to strengthen the building, to provide a base wall to which the finish siding can be nailed, to act as insulation, and in some cases to be a base for further insulation. Some of the common types of sheathing include, fiberboard, wood, and plywood.

FIBERBOARD SHEATHING

Fiberboard usually comes in 2 × 8 or 4 × 8 sheets which are tongue-and-grooved, and generally coated or impregnated with an asphalt material which increases water resistance. Thickness is normally 1/2 and 25/32 inches, and may be used where the stud spacing does not exceed 16 inches. Fiberboard sheathing should be nailed with 2-inch galvanized roofing nails or other type of non-corrosive nails. If the fiberboard is used as sheathing, most builders will use plywood at all corners (the thickness of the sheathing), to strengthen the walls, as shown in Fig. 1.

WOOD SHEATHING

Wood wall sheathing can be obtained in almost all widths, lengths, and grades. Generally, widths are from 6 to 12 inches, with lengths selected for economical use. Almost all solid wood wall sheathing used is 25/32 to 1 inch in thickness. This material may be nailed on horizontally or diagonally, as shown in Fig. 2. Wood sheathing is laid on tight, with all joints made over the

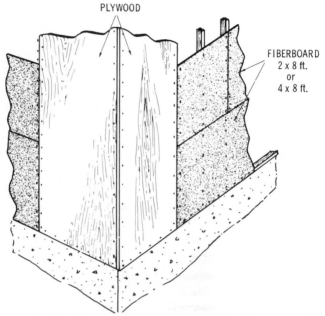

Fig. 1. Showing the method of using plywood on all corners as bracing when using fiberboard as exterior sheathing.

studs. If the sheathing is to be put on horizontally, it should be started at the foundation and worked toward the top. If the sheathing is installed diagonally, it should be started at the corners of the building and worked toward the center or middle.

Diagonal sheathing should be applied at a 45° angle. This method of sheathing adds greatly to the rigidity of the wall and eliminates the need for the corner bracing. It also provides an excellent tie to the sill plate when it is installed diagonally. There is more lumber waste than with horizontal sheathing because of the angle cut, and the application is somewhat more difficult. Fig. 3. shows the wrong way and the correct way of laying diagonal sheathing.

PLYWOOD SHEATHING

Plywood as a wall sheathing is highly recommended, because of its size, weight, and stability, plus the ease and rapidity of

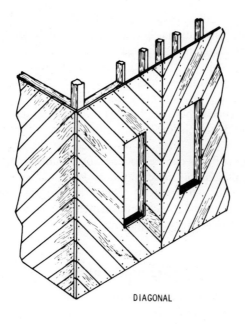

DIAGONAL

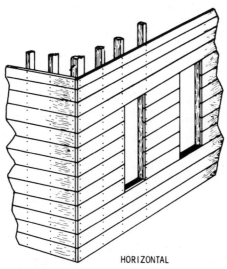

HORIZONTAL

Fig. 2. Two methods of nailing on wood sheathing.

97

application. It adds considerably more strength to the frame structure than the conventional horizontal or diagonal sheathing. When plywood sheathing is used, corner bracing can also be omitted. Large size panels effect a major saving in the time required for application and still provides a tight, draft-free installation that contributes a high insulation value to the walls. Minimum thickness of plywood wall sheathing is 5/16 inch for 16-inch stud spacing, and 3/8 inch for 24-inch stud spacing. The panels should be installed with the face grain parallel to the studs. However, a little more stiffness can be obtained by installing them across the studs, but this requires more cutting and fitting. Nail spacing should not be more than 6 inches on center at the edges of the panels and not more than 12 inches on center elsewhere. Joints should meet on the centerline of framing members.

SHEATHING PAPER

Sheathing paper should be used on frame structure when wood or plywood sheathing is used. It should be water resistant but not vapor resistant. It should be applied horizontally, starting at the bottom of the wall. Succeeding layers should lap about 4 inches, and lap over strips around openings. Strips about 6 inches wide should be installed behind all exterior trim or exterior openings.

WOOD SIDING

One of the materials most characteristic of the exteriors of American houses is wood siding. The essential properties required for wood siding are, good painting characteristics, easy working qualities, and freedom from warp. These properties are present to a high degree in the cedars, eastern white pine, sugar pine, western white pine, cypress, and redwood.

Material used for exterior siding should preferably be of a select grade, and should be free from knots, pitch pockets, and wavy edges. The moisture content at the time of application should be that which it would attain in service. This would be approximately 12 percent, except in the dry southwestern states, where the moisture content should average about 9 percent.

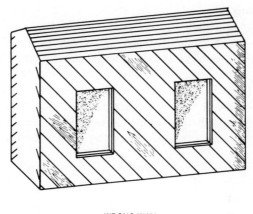

WRONG WAY

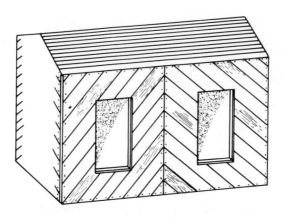

CORRECT WAY

Fig. 3. Wrong and correct way of laying sheathing.

Bevel Siding

Plain bevel siding, as shown in Fig. 4, is made in nominal 4-, 5-, and 6-inch widths with 7/16-inch butts, 6-, 8-, and 10-inch widths with 9/16 and 11/16-inch butts. Bevel siding is generally furnished in random lengths varying from 4 to 20 feet in length.

Drop Siding

Drop siding is generally 3/4 inch thick, and is made in a variety of patterns with either matched or shiplap edges. Fig. 5 shows three common patterns of drop siding that are applied horizontally. Fig. 5A may be applied vertically, for example at the gable ends of a house. Drop siding was designed to be applied directly to the studs, and it thereby serves as sheathing and exterior wall covering. It is widely used in this manner in farm structures, such as sheds and garages in all parts of the country. When used over or when

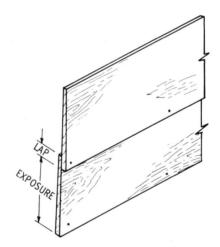

Fig. 4. Illustrating bevel siding.

in contact with other material, such as sheathing or sheathing paper, water may work through the joints and be held between the sheathing and the siding. This sets up a condition conducive to paint failure and decay. Such problems can be avoided when the side walls are protected by a good roof overhang.

Square-Edge Siding

Square-edge or clapboard siding made of 25/32-inch board is occasionally selected for architectural effects. In this case, wide boards are generally used. Some of this siding is also beveled on the back at the top to allow the boards to lie rather close to the sheathing, thus providing a solid nailing surface.

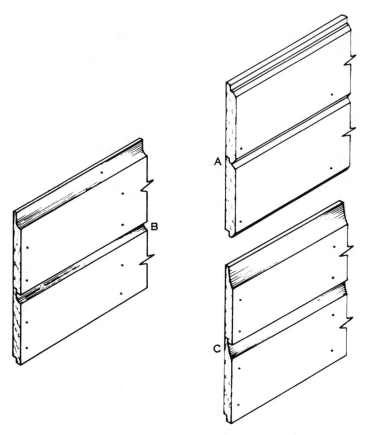

Fig. 5. Types of drop siding (A) V-rustic, (B) drop, (C) rustic drop.

Vertical Siding

Vertical siding is commonly used on the gable ends of a house, over entrances, and sometimes for large wall areas. The type used may be plain-surfaced matched boards, patterned matched boards or square-edge boards covered at the joint with a batten strip. Matched vertical siding should preferably not be more than 8 inches wide and should have 2 eight-penny nails not more than 4 feet apart. Backer blocks should be placed between studs to provide a good nailing base. The bottom of the boards should be undercut to form a water drip.

101

Batten-type siding is often used with wide square-edged boards which, because of their width, are subjected to considerable expansion and contraction. The batten strips used to cover the joints should be nailed to only one siding board so the adjacent board can swell and shrink without splitting the boards or the batten strip.

Plywood Siding

Plywood is often used in gable ends, sometimes around windows and porches, and occasionally as an overall exterior wall covering. The sheets are made either plain or with irregularly cut striations. It can be applied horizontally or vertically. The joints can be moulded batten, V-grooves, or flush. Sometimes it is installed as lap siding. Plywood siding should be of exterior grade, since houses are often built with little overhang of the roof, particu-

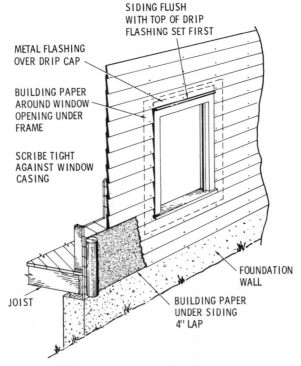

Fig. 6. Installation of bevel siding.

larly on the gable end. This permits rainwater to run down freely over the face of the siding. For unsheathed walls, the following thicknesses are suggested:

Minimum thickness	Maximum stud space
3/8 inch	16 inches on center
1/2 inch	20 inches on center
5/8 inch	24 inches on center

Treated Siding

Houses are often built with little or no overhang of the roof, particularly on the gable ends. This permits rain water to run down freely over the face of the siding. Under such conditions water may work up under the laps in bevel siding or through joints in drop siding by capillary action, and provide a source of moisture that may cause paint blisters or peeling.

A generous application of a water repellent preservative to the back of the siding will be quite effective in reducing capillary action with bevel siding. In drop siding, the treatment would be applied to the matching edges. Dipping the siding in the water repellent would be still more effective. The water repellent should be applied to all end cuts, at butt points, and where the siding meets door and window trim.

INSTALLATION OF SIDING

The spacing for siding should be carefully laid out before the first board is applied. The bottom of the board that passes over the top of the first-floor windows should coincide with the top of the window cap, as shown in Fig. 6. To determine the maximum board spacing or exposure, deduct the minimum lap from the overall width of the siding. The number of board spaces between the top of the window and the bottom of the first course at the foundation wall should be such that the maximum exposure will not be exceeded. This may mean that the boards will have less than the maximum exposure.

Siding starts with the bottom course of boards at the foundation, as shown in Fig. 7. Sometimes the siding is started on a water table, which is a projecting member at the top of the foundation to throw off water, as shown in Fig. 8. Each succeeding

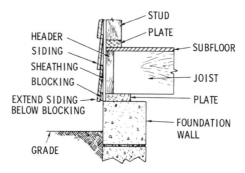

Fig. 7. *Installation of the first or bottom course.*

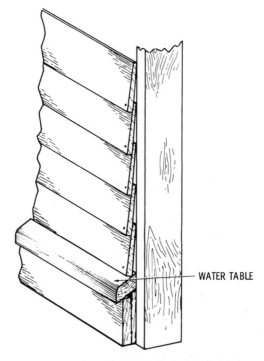

Fig. 8. *Illustration of the water table which is sometimes used.*

course overlaps the upper edge of the lower course. The minimum head lap is 1 inch for 4 and 6 inch widths, and 1-1/4 inch lap for widths over 6 inches. The joints between boards in adjacent courses should be staggered as much as possible. Butt joints should always be made on a stud, or where boards butt against window and door casings and corner boards. The siding should be carefully fitted and be in close contact with the member or adjacent pieces. Some carpenters fit the boards so tight that they have to spring the boards in place, which assures a tight joint. Loose-fitting joints allow water to get behind the siding and thereby causes paint deterioration around the joints, and also sets up conditions conducive to decay at the ends of the siding.

Types of Nails

Nails cost very little compared to the cost of siding and labor, but the use of good nails is important. It is poor economy to buy siding that will last for years and then use nails that will rust badly within a few years. Rust resistant nails will hold the siding permanently and will not disfigure light-colored paint surfaces.

There are two types of nails commonly used with siding, one having a small head and the other a slightly larger head. The small-head casing nail is set (driven with a nailset) about 1/16 inch below the surface of the siding. The hole is filled with putty after the prime coat of paint is applied. The large-head nail is driven flush with the face of the siding, with the head being later covered with paint. Ordinary steel wire nails tend to rust in a short time and cause a disfiguring stain on the face of the siding. In some cases, the small-head nail will show rust spots through the putty and paint. Noncorrosive-type nails (galvanized, aluminum, and stainless steel) that will not cause rust stains are readily available.

Bevel siding should be face nailed to each stud with noncorrosive nails, the size depending upon the thickness of the siding and the type of sheathing used. The nails are generally placed about 1/2 inch above the butt edge, in which case it passes through the upper edge of the lower course of siding. Another method recommended for bevel siding by most associations representing siding manufacturers, is to drive the nails through the siding just above the lap so that the nail misses the thin edge of the piece of siding underneath. The latter method permits expansion and contraction of the siding board with seasonal changes in moisture content.

Corner Treatment

The method of finishing the wood siding at the exterior corners is influenced somewhat by the overall house design. Corner boards are appropriate to some designs, and mitered joints to others. Wood siding is commonly joined at the exterior corners by corner boards, mitered corners, or by metal corners.

Corner Boards—Corner boards, as shown in Fig. 9, are used with bevel or drop siding and are generally made of nominal 1 or 1-1/4 inch material, depending upon the thickness of the siding. It may be either plain or moulded, depending on the architectural treatment of the house. The corner boards may be

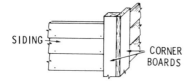

SIDING

CORNER BOARDS

Fig. 9. Corner treatment for bevel siding using the corner board.

applied vertically against the sheathing, with the siding fitting tightly against the narrow edge of the corner board. The joints between the siding and the corner boards and trim should be caulked or treated with a water repellent. Corner boards, and trim around windows and doors, are sometimes applied over the siding, a method that minimizes the entrance of water into the ends of the siding.

Mitered Corners—Mitered corners, such as shown in Fig. 10, must fit tightly and smoothly for the full depth of the miter. To

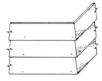

Fig. 10. Illustrating the mitered corner treatment.

maintain a tight fit at the miter, it is important that the siding is properly seasoned before delivery, and is stored at the site so as to be protected from rain. The ends should be set in white lead when the siding is applied, and the exposed faces should be primed immediately after it is applied. At interior corners, shown in Fig. 11, the siding is butted against a corner strip of nominal 1 or 1-1/4 inch material, depending upon the thickness of the siding.

Metal Corners—Metal corners, as shown in Fig. 12, are made of 8-gauge metals, such as aluminum and galvanized iron. They are used with bevel siding as a substitute for mitered corners, and can be purchased at most lumber yards. The application of metal corners takes less skill than is required to make good mitered corners, or to fit the siding to a corner board. Metal corners should always be set in white lead paint.

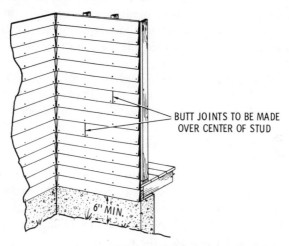

BUTT JOINTS TO BE MADE
OVER CENTER OF STUD

6" MIN.

Fig. 11. Illustrating the construction of an interior corner using bevel siding.

METAL SIDING

The metal most popular at the present time is aluminum. It is installed over most types of sheathing with an aluminum building paper (for insulation) nailed on between the sheathing and siding. Its most attractive characteristic is the long-lasting finish obtained

107

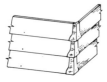

Fig. 12. Corner treatment for bevel siding using the corner metal caps.

on the prefinished product. The cost of painting and maintenance has made this type of siding doubly attractive. Aluminum siding can be installed over old siding that has cracked and weathered, or where paint will not hold up.

SUMMARY

Sheathing is nailed directly to the framework of the building. The purpose of sheathing is to strengthen the structure, to provide a nailing base for outside weatherboarding, and to act as insulation. Various types of sheathing include fiberboard, wood, and plywood.

Fiberboard generally is furnished in 2 × 8 or 4 × 8 foot sheets and usually coated with an asphalt material to make waterproof. When fiberboard sheathing is used, most builders will use plywood at all corners to strengthen the walls. Fiberboard is normally ½ or 25/32 inches thick and generally tongue-and-grooved.

Wood sheathing is generally any size from 1 × 6 to 1 × 12 inches in width. The material may be installed horizontally or diagonally with all joints made over a stud. Diagonal sheathing should be applied at a 45° angle. This adds greatly to the rigidity of the walls and eliminates the need for corner bracing. More lumber waste is realized than when applying horizontal sheathing, but an excellent tie to the sill plate is accomplished when installed diagonally.

One of the most popular exterior wall finishes of American houses is wood siding. Various types or styles include bevel, drop, square-edge, and vertical siding. A number of methods are used as a corner treatment when using wood bevel siding. Some corners are designed to use a vertical corner board, which is generally 1-

or 1¼-inch material. Mitered corners are sometimes used, or the same effect can be obtained by using metal corners.

REVIEW QUESTIONS

1. What is fiberboard and how is it used as sheathing?
2. What are some advantages in using wood sheathing placed diagonally?
3. Name the various styles of wood siding.
4. How are corners on wood siding treated?
5. What is a water table?

CHAPTER 7

Stairs

All carpenters who have tried to build stairs have found it (like boat building) to be an art in itself. This chapter is not intended to discourage the carpenter, but to impress him with the fact that unless he first masters the principle of stair layout, he will have many difficulties in the construction. Although stair building is a branch of mill work, the carpenter should know the principles of simple stair layout and construction, because he is often called upon to construct porch steps, basement and attic stairs, and sometimes the main stairs. In order to follow the instructions intelligently, the carpenter should be familiar with the terms and names of parts used in stair building.

STAIR CONSTRUCTION

Stairways should be designed, arranged, and installed so as to afford safety, adequate headroom, and space for the passage of furniture. In general, there are two types of stairs in a house—those serving as principal stairs, and those used as service stairs. The principle stairs are designed to provide ease and comfort, and are often made a feature of design, while the service stairs leading to the basement or attic are usually somewhat steeper and constructed of less expensive materials.

Stairs may be built in place, or they may be built as units in the shop and set in place. Both have their advantages and disadvantages, and custom varies with locality. Stairways may have

111

a straight continuous run, with or without an intermediate platform, or they may consist of two or more runs at angles to each other. In the best and safest practice, a platform is introduced at the angle, but the turn may be made by radiating risers called *winders*. Nonwinder stairways are most frequently encountered in residential planning, because winder stairways represent a condition generally regarded as undesirable. However, use of winders is sometimes necessary because of cramped space. In such instances, winders should be adjusted to replace landings so that the width of the tread 18 inches from the narrow converging end will not be less than the tread width on the straight run.

RATIO OF RISER TO TREAD

There is a definite relation between the height of a riser and the width of a tread, and all stairs should be laid out to conform to the well established rules governing these relations. If the combination of run and rise is too great, the steps are tiring, placing a strain on the leg muscles and on the heart. If the steps are too short, the foot may kick the leg riser at each step and an attempt to shorten the stride may be tiring. Experience has proved that a riser 7 to 7-1/2 inches high, with appropriate tread, combines both comfort and safety, and these limits therefore determine the standard height of risers commonly used for principal stairs. Service stairs may be narrow and steeper than the principal stairs, and are often unduly so, but it is well not to exceed 8 inches for the risers.

As the height of the riser is increased, the width of the tread must be decreased for comfortable results. A very good ratio is provided by either of the following rules, which are exclusive of the nosing:

1. Tread plus twice the riser equals 25.
2. Tread multiplied by the riser equals 75.

A riser of 7-1/2 inches would, therefore, require a tread of 10 inches, and a riser of 6-1/2 inches would require a tread of 12 inches wide. Treads are rarely made less than 9 inches or more than 12 inches wide. The treads of main stairs should be made of prefinished hardwood.

DESIGN OF STAIRS

The location and the width of a stairway (together with the platforms) having been determined, the next step is to fix the height of the riser and width of the tread. After a suitable height of riser is chosen, the exact distance between the finish floors of the two stories under consideration is divided by the riser height. If the answer is an *even* number, the number of risers is thereby determined. It very often happens, that the results is *uneven,* in which case the story height is divided by the whole number next above or below the quotient. The result of this division gives the height of the riser. The tread is then proportioned by one of the rules for ratio of riser to tread.

Assume that the total height from one floor to the top of the next floor is 9′ 6″, or 114 inches, and that the riser is to be approximately 7-1/2 inches. The 114 inches would be divided by 7-1/2 inches, which would give 15-1/5″ risers. However, the number of risers must be an *equal or whole* number. Since the nearest whole number is 15, it may be assumed that there are to be 15 risers, in which case 114 divided by 15 equals 7.6 inches, or approximately 7-9/16 inches for the height of each riser. To determine the width of the tread, multiply the height of the riser by 2 (2 × 7-9/16 = 15-1/8), and deduct 25 (25 − 15/18 = 9-7/8 inches).

The headroom is the vertical distance from the top of the tread to the underside of the flight or ceiling above, as shown in Fig. 1. Although it varies with the steepness of the stairs, the minimum allowed would be 6′ 8″.

FRAMING OF STAIR WELL

When large openings are made in the floor, such as for a stair well, one or more joists must be cut. The location in the floor has a direct bearing on the method of framing the joists.

The principles explained in Chapter 8 of *Carpenters and Builders Guide No. 3,* may be referred to in considering the framing around openings in floors for stairways. The framing members

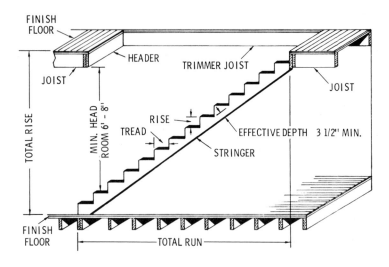

Fig. 1. Stairway design.

around these openings are generally of the same depth as the joists. Fig. 2 shows the typical framing around a stair well and landing.

The headers are the short beams at right angle to the regular joists at the end of the floor opening. They are doubled and support the ends of the joists that have been cut off. Trimmer joists are at the sides of the floor opening, and run parallel to the regular joists. They are also doubled and support the ends of the headers. Tail joist are joists that run from the headers to the bearing partition.

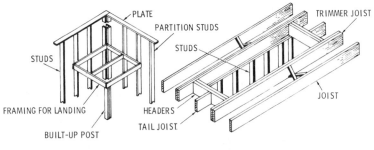

(A) For landings. (B) For straight run stair well.

Fig. 2. Framing of stairways.

114

STRINGERS OR CARRIAGES

The treads and risers are supported upon stringers or carriages that are solidly fixed in place, and are level and true on the framework of the building. The stringers may be cut or ploughed to fit the outline of the tread and risers. The third stringer should be

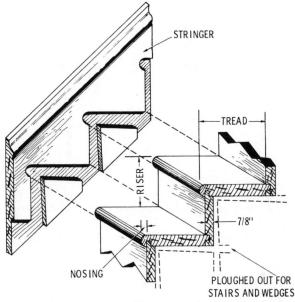

Fig. 3. Showing the housing in the stringer board for the tread and riser.

installed in the middle of the stairs when the treads are less than 1-1/8 inches thick and the stairs are more than 2′ 6″ wide. In some cases, rough stringers are used during the construction period. These have rough treads nailed across the stringers for the convenience of workmen until the wall finish is applied. There are several forms of stringers classed according to the method of attaching the risers and treads. These different types are, *cleated, cut, built-up,* and *rabbetted.*

When the wall finish is complete, the finish stairs are erected or built in place. This work is generally done by a stair builder, who often operates as a member of a separate specialized craft. The wall stringer may be ploughed out, or rabbetted, as shown in

Fig. 3, to the exact profile of the tread, riser, and nosing, with sufficient space at the back to take the wedges. The top of the

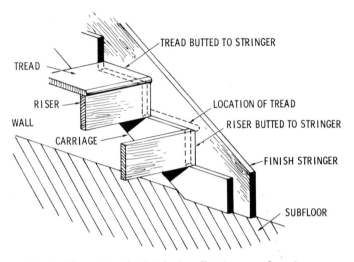

Fig. 4. Illustrating the finished wall stringer and carriage.

riser is tongued into the front of the tread and into the bottom of the next riser. The wall stringer is spiked to the inside of the wall, and the treads and risers are fitted together and forced into the wall stringer nosing, where they are set tight by driving and gluing the wood wedges behind them. The wall stringer shows above the profiles of the tread and riser as a finish against the wall and is often made continuous with the baseboard of the upper and lower landing. If the outside stringer is an open stringer, it is cut out to fit the risers and treads, and nailed against the outside carriage. The edges of the riser are mitered with the corresponding edges of the stringer, and the nosing of the tread is returned upon its outside edge along the face of the stringer. Another method would be to butt the stringer to the riser and cover the joint with an inexpensive stair bracket.

Fig. 4 shows a finish stringer nailed in position on the wall, and the rough carriage nailed in place against the stringer. If there are walls on both sides of the staircase, the other stringer and carriage would be located in the same way. The risers are nailed to the riser cuts of the carriage on each side and butt against each

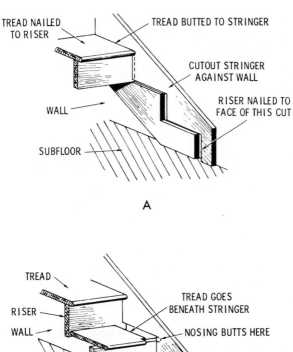

A

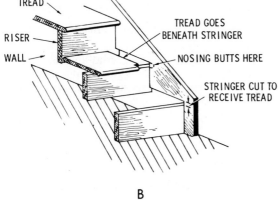

B

Fig. 5. Showing the stringers and treads.

side of the stringer. The treads are nailed to the tread cuts of the carriage and butt against the stringer. This is the least expensive of the types described and perhaps the best construction to use when the treads and risers are to be nailed to the carriages.

Another method of fitting the treads and risers to the wall stringers is shown in Fig. 5A. The stringers are laid out with the same rise and run as the stair carriages, but they are cut out in

117

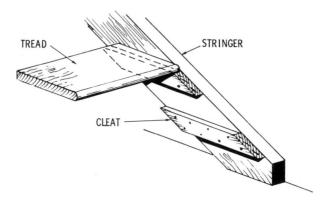

Fig. 6. Showing the cleats stringer used in basement stairs.

reverse. The risers are butted and nailed to the riser cuts of the wall stringers, and the assembled stringers and risers are laid over the carriage. Sometimes the treads are allowed to run underneath the tread cut of the stringer. This makes it necessary to notch the tread at the nosing to fit around the stringer, as shown in Fig. 5B.

Another form of stringer is the cut-and-mitered type. This is a form of open stringer in which the ends of the risers are mitered against the vertical portion of the stringer. This construction is shown in Fig. 8, and is used when the outside stringer is to be finished and must blend with the rest of the casing or skirting board. A moulding is installed on the edge of the tread and carried around to the side, making an overlap as shown in Fig. 9.

BASEMENT STAIRS

Basement stairs may be built either with or without riser boards. Cutout stringers are probably the most widely used support for the treads, but the tread may be fastened to the stringers by cleats, as shown in Fig. 6. Fig. 7 shows two methods of terminating basement stairs at the floor line.

NEWELS AND HANDRAILS

All stairways should have a handrail from floor to floor. For closed stairways, the rail is attached to the wall with suitable metal

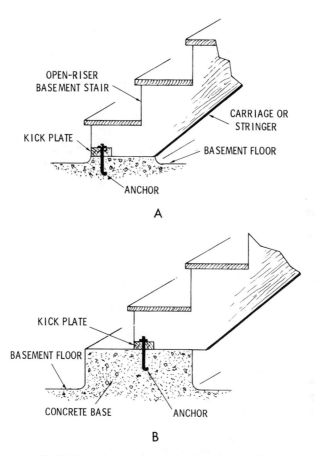

OPEN-RISER
BASEMENT STAIR

CARRIAGE OR
STRINGER

KICK PLATE

BASEMENT FLOOR

ANCHOR

A

KICK PLATE

BASEMENT FLOOR

CONCRETE BASE

ANCHOR

B

Fig. 7. Basement stair termination at floor line.

brackets. The rails should be set 2′ 8″ above the tread at the riser line. Handrails and balusters are used for open stairs and for open spaces around stairs. The handrail ends against the newel post, as shown in Fig. 10.

Stairs should be laid out so that stock parts may be used for newels, rails, balusters, goosenecks, and turnouts. These parts are a matter of design and appearance, so they may be very plain or elaborate, but they should be in keeping with the style of the house. The balusters are doweled or dovetailed into the treads and, in some cases, are covered by a return nosing. Newel posts should

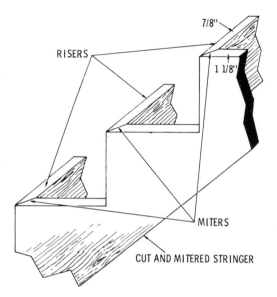

RISERS

7/8"

1 1/8"

MITERS

CUT AND MITERED STRINGER

Fig. 8. Cut and mitered stringer.

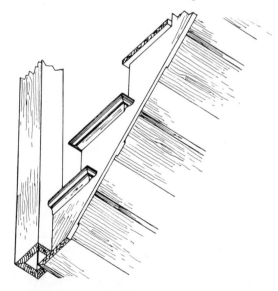

Fig. 9. Illustrating the use of moulding on the edge of treads.

be firmly anchored, and where half-newels are attached to a wall, blocking should be provided at the time the wall is framed.

DISAPPEARING STAIRS

Where attics are used primarily for storage, and where space for a fixed stairway is not available, hinged or disappearing stairs are often used. Such stairways may be purchased ready to install. They operate through an opening in the ceiling of a hall and swing up into the attic space, out of the way when not in use. Where such stairs are to be provided, the attic floor should be designed for regular floor loading.

EXTERIOR STAIRS

Proportioning of risers and treads in laying out porch steps or approaches to terraces should be as carefully considered as the design of interior stairways. Similar riser-to-tread ratios can be

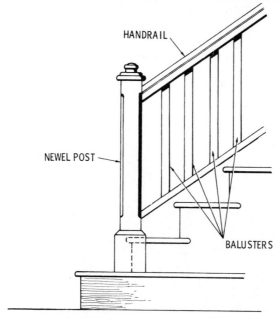

Fig. 10. Illustrating the newel post, balusters, and handrail.

121

used, however. The riser used in principal exterior steps should be between 6 and 7 inches. The need for a good support or foundation for outside steps is often overlooked. Where wood steps are used, the bottom step should be set in concrete. Where the steps are located over back fill or disturbed ground, the foundation should be carried down to undisturbed ground. Fig. 11 shows the foundation and details of the step treads, handrail, and stringer,

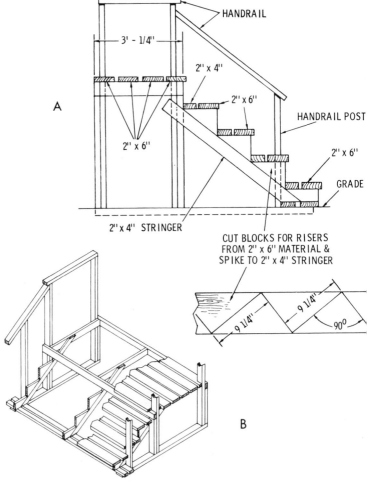

Fig. 11. Outside step construction.

and the method of installing them. This type of step is most common in field construction and outside porch steps. The material generally used for this type of stair construction are 2 × 4's and 2 × 6'se.

Glossary of Stair Terms

The terms generally used in stair design, may be defined as follows.

Balusters—The vertical members supporting the handrail on open stairs (Fig. 12).

Carriage—The rough timber supporting the treads and risers of wood stairs, sometimes referred to as the string or stringer, as shown in Fig. 13.

Fig. 12. Illustrating the baluster which supports the handrail.

Circular Stairs—A staircase with steps planned in a circle, all the steps being winders (Fig 14).

Flight of Stairs—The series of steps leading from one landing to another.

Front String—The string of that side of the stairs over which the hand rail is placed.

Fillet—A band nailed to the face of a front string below the curve and extending the width of a tread.

Flyers—Steps in a flight of stairs parallel to each other.

Half-Space—The interval between two flights of steps in a staircase.

Handrail—The top finishing piece on the railing intended to be grasped by the hand in ascending and descending. For closed stairs where there is no railing, the handrail is attached to the wall with brackets. Various forms of hand rails are shown in Fig. 15.

Housing—The notches in the string board of a stair for the reception of steps.

Landing—The floor at the top or bottom of each story where the flight ends or begins.

Newel—The main post of the railing at the start of the stairs and the stiffening posts at the angles and platform.

Nosing—The projection of tread beyond the face of the riser (Fig. 16).

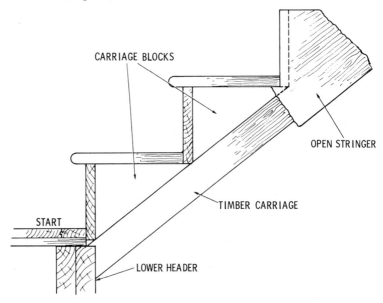

Fig. 13. Illustrating the carriage blocks connected to a stair stringer.

Rise—The vertical distance between the treads or for the entire stairs.

Riser—The board forming the vertical portion of the front of the step, as shown in Fig. 17.

Run—The total length of stairs including the platform.

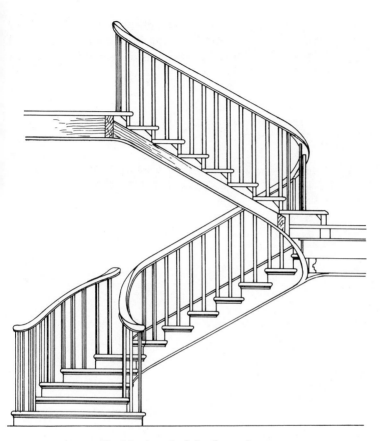

Fig. 14. A typical circular staircase.

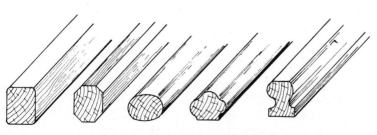

Fig. 15. Various forms of handrails.

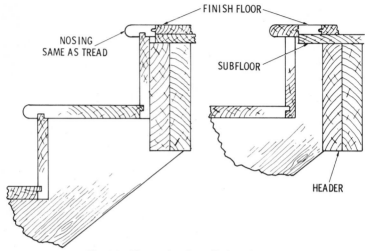

Fig. 16. The nosing installed to the tread.

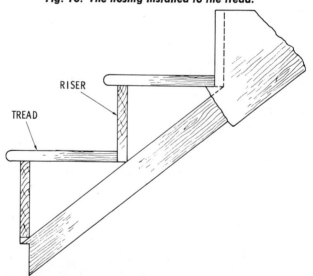

Fig. 17. Illustrating the tread and riser.

Stairs—The steps used to ascend and descend from one story to another.

Staircase—The whole set of stairs with the side members supporting the steps.

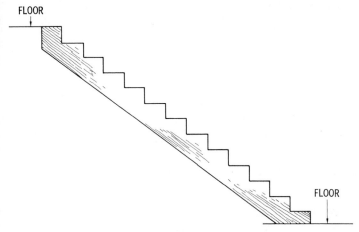

Fig. 18. Illustrating the stair stringer.

Straight Flight of Stairs—One having the steps parallel and at right angles to the strings.

String or Stringer—One of the inclined sides of a stair supporting the tread and riser. Also, a similar member, whether a support or not, such as finish stock placed exterior to the carriage on open stairs, and next to the walls on closed stairs, to give finish to the staircase. *Open stringers,* both rough and finish stock, are cut to follow the lines of the treads and risers. *Closed stringers* have parallel sides, with the risers and treads being housed into them (Fig. 18).

Tread—The horizontal face of a step, as shown in Fig. 17.

Winders—The radiating or wedge-shaped treads at the turn of a stairway.

SUMMARY

Stairways should always be designed, arranged, and installed so as to afford safety, adequate headroom, and space for passage of furniture. Stairs may be built in place, or they may be built as a complete unit in the shop and set in place.

There is a definite relation between the height of a riser and the width of a tread. If the steps are too short, the foot may kick the

127

leg riser at each step and an attempt to shorten the stride may be tiring. As the height of the riser is increased, the width of the tread must be decreased for comfortable results.

When openings are made in a floor, such as for a stair well, headers and trimmer joists must be used to strengthen the floor around the opening. Stringers, which are the supports for the tread and riser, are installed between floor levels. There are several forms of stringers classed according to the method of attaching the risers and treads. The different types are cleated, cut, built-up, and rabbetted.

REVIEW QUESTIONS

1. What are stringers, risers, and treads?
2. Name the four types of stringers.
3. When is a center or third stringer used?
4. How is the rise figured when designing a stairway?
5. What is the carriage of a stairs?

CHAPTER 8

Flooring

After the foundation, sills, and floor joists have been constructed, the subfloor is laid diagonally on the joists. The floor joist forms a framework for the subfloor. This floor is called the rough floor, or subfloor, and may be visioned as a large platform covering the entire width and length of the building. Two layers or coverings of flooring material (subflooring and finished flooring) are placed on the joists. Boards 6, 8, or 12 inches wide, and generally 1 inch in thickness, are used. Plywood is also used in some cases, because of its size, weight, and stability, plus the time and labor saved in application. It will add considerable more strength to the floor since the weight is distributed over a wider area. Fig. 1 shows the method of laying a subfloor.

It may be laid before or after the walls are framed, preferably before, so it can be used as a floor to work on while framing the walls. The subflooring will also give protection against the weather for tools and material stored in the basement.

SUBFLOORING

Subflooring should consist of square edge or tongue-and-groove boards, or plywood. Tongue-and-groove end-matched boards may be used, but they should be applied so that each board will bear on at least two joists, and so that there will be no two adjoining boards with end joints occurring between the same pair of joists. Subflooring is nailed to each joist with two eightpenny nails for

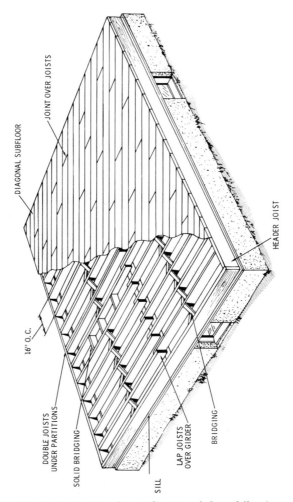

Fig. 1. Illustrating the application of the subflooring.

widths under 8 inches and with three nails for over 8 inch width. The subflooring may be applied either diagonally or at right angles to the joists. When the subfloor is placed at right angles to the joists, the finish floor should be laid at right angles to the subflooring. Diagonal subflooring permits the finish floor to be laid either parallel or perpendicular to the joists.

The joist spacing should not exceed 16 inches on center when finish flooring is laid parallel to the joists or when parquet finish flooring is used, and should not exceed 24 inches on center when finish flooring at least 25/32 inch thick is at right angles to the joists. Where balloon framing is used, blocking should be installed between the ends of the joists at the wall for nailing the ends of diagonal subfloor boards. In areas where rain may occur during construction, square-edge boards should be laid with open joints for drainage. Tongue-and-groove boards should have holes drilled at suitable intervals to allow runoff of rain water.

Table 1 shows the thickness of the plywood and joist spacings as suggested by the Federal Housing Administration for plywood subfloor when used as a base for wood finish floors, linoleum, composition, rubber, or ceramic tile.

When used as a base for parquet wood finish flooring less than 25/32 inch thick, or linoleum, composition, rubber, or ceramic tile, install solid blocking under all edges at right angles to the floor joists. Nail securely to the joists and blocking with nails 6 inches on center at the edges and 10 inches on center at the intermediate framing members. When used for leveling purposes over other subflooring, the minimum thickness is 1/4 inch three-ply.

FLOOR COVERINGS

The term "finished flooring" applies to the material used as the final wearing surface that is applied to a floor. There are many of these materials, each one having properties suited to a particular usage. Of these properties, durability and ease of cleaning are essential in all cases. Specific service requirements may call for special properties, such as resistance to hard wear in storehouses and on loading platforms; comfort to users in offices and shops; and attractive appearance, which is always desirable in residences.

There is a wide selection of material that may be used for flooring. Hardwoods and softwoods are available as strip flooring in a variety of widths and thicknesses, as well as random-width planks, parquetry, and block flooring. Other materials include plain and inlaid linoleum, asphalt, rubber, and ceramic tile.

131

Table 1. Plywood Thickness and Joists Spacing

Minimum thickness of five-ply subfloor	Medium thickness of finish flooring	Maximum joist spacing
†½ inch	25/32 inch wood laid at right angles to joists	24 inches
†½ inch	25/32 inch wood laid parallel to joists	20 inches
½ inch	25/32 inch wood laid at right angles to joists	20 inches
½ inch	Less than 25/32 inch wood or other finish	‡16 inches
†⅝ inch	Less than 25/32 inch wood or other finish	‡20 inches
†¾ inch	Less than 25/32 inch wood or other finish	‡24 inches

† Installed with outer plies of subflooring at right angles to joists.
‡ Wood strip flooring, 25/32 inch thick or less, may be applied in either direction.

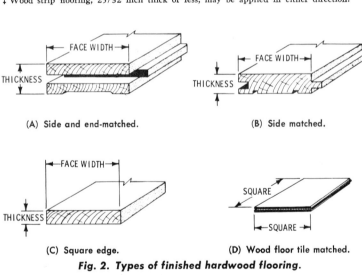

(A) Side and end-matched. (B) Side matched.

(C) Square edge. (D) Wood floor tile matched.

Fig. 2. Types of finished hardwood flooring.

Wood Strip Flooring

Softwoods most commonly used for flooring are southern yellow pine, Douglas fir, redwood, western larch, and western hemlock. It is customary to divide the softwoods into two classes:

132

1. Vertical or edge grain.
2. Flat grain.

Each class is separated into select and common grades. The select grades designated as "B and better" grades, and sometimes the "C" grade, are used when the purpose is to stain, varnish, or wax the floor. The "C" grade is well suited for floors to be stained dark or painted, and lower grades are for rough usage when covered with carpeting. Softwood flooring is manufactured in several widths. In some places, the 2-1/2 inch width is preferred, while in others, the 3-1/2 inch width is more popular. Softwood flooring has tongue-and-groove edges, and may be hollow backed or grooved. Vertical-grain flooring stands up better than flat-grain under hard usage.

Hardwoods most commonly used for flooring are red and white oak, hard maple, beech, and birch. Maple, beech, and birch comes in several grades, such as *first, second* and *third*. Other hardwoods that are manufactured into flooring, although not commonly used, are walnut, cherry, ash, hickory, pecan, sweetgum, and sycamore. Hardwood flooring is manufactured in a variety of widths and thicknesses, some of which are referred to as standard patterns, others as special patterns. The widely used standard patterns consist of relatively narrow strips laid lengthwise in a room, as shown in Fig. 2. The most widely used standard pattern is 25/32 inch thick and has a face width of 2 1/4 inches. One edge has a tongue and the other end has a groove, and the ends are similarly matched. The strips are random lengths, varying from 1 to 16 feet in length. The number of short pieces will depend on the grade used. Similar patterns of flooring are available in thicknesses of 15/32 and 11/32 inch, with a face width of 1-1/2 inches, and with square edges and a flat back.

The flooring is generally hollow backed. The top face is slightly wider than the bottom, so that the strips are driven tight together at the top side but the bottom edges are slightly open. The tongue should fit snugly in the groove to eliminate squeaks in the floor.

Another pattern of flooring used to a limited degree is 3/8 inch thick with a face width of 1-1/2 and 2 inches, with square edges and a flat back. Fig. 2D shows a type of wood floor tile commonly known as parquetry.

INSTALLATION OF WOOD STRIP FLOORING

Flooring should be laid after plastering and other interior wall and ceiling finish is completed, after windows and exterior doors are in place, and after most of the interior trim is installed. When wood floors are used, the subfloor should be clean and level, and should be covered with a deadening felt or heavy building paper, as shown in Fig. 3. This felt or building papers will stop a certain amount of dust, and will somewhat deaden the sound. Where a crawl space is used, it will increase the warmth of the floor by preventing air infiltration. The location of the joists should be chalklined on the paper as a guide for nailing.

Strip flooring should be laid crosswise of the floor joist, and it looks best when the floor is laid lengthwise in a rectangular room. Since joists generally span the short way in a living room, that room establishes the direction for flooring in other rooms. Flooring should be delivered only during dry weather, and should be stored in the warmest and driest place available in the house. The recommended average moisture content for flooring at the time of installation should be between 6 and 10 percent. Moisture absorbed after delivery to the house site is one of the most common causes of open joints between floor strips. It will show up several months after the floor has been laid.

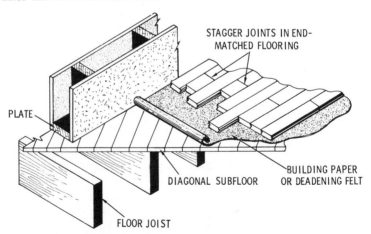

Fig. 3. Application of strip flooring showing the use of deadening felt or heavy building paper.

Floor squeaks are caused by movement of one board against another. Such movement may occur because the floor joists are too light and are not held down tightly, tongue fitting too loose in the grooves, or because of poor nailing. Adequate nailing is one of the most important means of minimizing squeaks. When it is possible to nail the finish floor through the subfloor into the joist, a much better job is obtained than if the finish floor is nailed only to the subfloor. Various types of nails are used in nailing various thicknesses of flooring. For 25/32-inch flooring, it is best to use eightpenny steel cut flooring nails; for 1/2-inch flooring, sixpenny nails should be used. Other types of nails have been developed in recent years for nailing of flooring, among these being the annularly-grooved and spirally-grooved nails. In using these nails, it is well to check with the floor manufacturer's recommendation as to size and diameter for a specific use. Fig. 4 shows the method of nailing the first strip of flooring. The nail is driven straight down through the board at the groove edge. The nails should be driven into the joist and near enough to the edge so that they will be covered by the base or shoe moulding. The first strip of flooring can be nailed through the tongue.

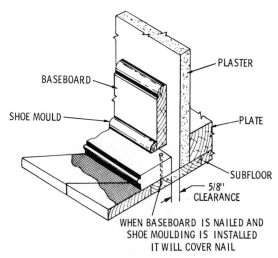

Fig. 4. Method of laying the first strips of wood flooring.

135

Fig. 5 shows the nail driven in at an angle of between 45° and 50° where the tongue adjoins the shoulder. Do not try to drive the nail down with a hammer, as the wood may be easily struck and damaged. Instead use a nail set to finish off the driving. Fig. 6 shows the position of the nail set commonly used for the final driving. In order to avoid splitting the wood, it is sometimes necessary to pre-drill the holes through the tongue. This will also help to drive the nail easily into the joist. For the second course of flooring, select a piece so that the butt joints will be well separated from those in the first course. For floors to be covered with rugs, the long lengths could be used at the sides of the room and the short lengths in the center where they will be covered.

Each board should be driven up tightly, but do not strike the tongue with the hammer, as this may crush the wood. Use a piece of scrap flooring for a driving block. Crooked pieces may require wedging to force them into alignment. This is necessary in order that the last piece of flooring will be parallel to the base-

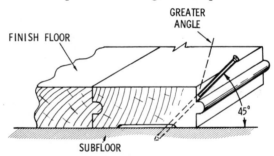

Fig. 5. Nailing method for strip wood flooring.

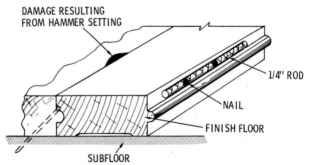

Fig. 6. Suggested method for setting nails in flooring.

board. If the room is not square, it may be necessary to start the alignment at an early stage.

SOUND-PROOF FLOORS

One of the most effective sound resistance floors is called a *floating* floor. The upper or finish floor is constructed on 2×2 joists actually floating on glass wool mats, as shown in Fig. 7. There should be absolutely no mechanical connection through the glass wool mat, not even a nail to either the subfloor or to the wall.

WOOD-TILE FLOORING

Flooring manufacturers have developed a wide variety of special patterns of flooring, sometimes called floor tile, and sometimes re-

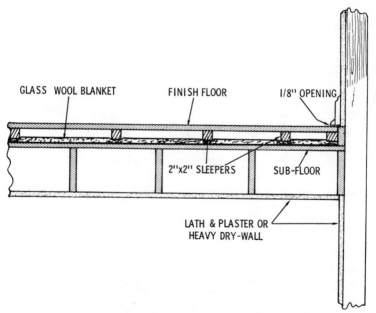

GLASS WOOL BLANKET   FINISH FLOOR   1/8" OPENING

2"x2" SLEEPERS   SUB-FLOOR

LATH & PLASTER OR
HEAVY DRY-WALL

Fig. 7. Illustrating a sound resistant floor.

ferred to as parquetry flooring, that can be used over wood subfloors or concrete slabs. One common type of floor tile is a block 9 inches square and 13/16 inches thick which is made up of several individual strips of flooring held together with glue and splines. Two edges have a tongue and the opposite edges are grooved. Numerous other sizes and thickness are available. In laying the floor, the direction of the blocks is alternated to create a checkerboard effect. The manufacturer supplies instructions for laying their tile, and it is advisable to follow them carefully. When the tiles are used over a concrete slab, a vapor barrier should be used, as shown in Fig. 8. The slab should be level, and thoroughly cured and dry before the wood tile is laid.

BASE FOR LINOLEUM, ASPHALT, AND RUBBER TILE

Linoleum, asphalt, and rubber floor tile may be laid directly on 25/32-inch tongue-and-groove wood flooring strips, with a maximum width of 3-1/4 inches. Plywood subflooring (1/2 thick) may be used if joists are spaced 16 inches on center. Where these finish floors are used for the floor covering in some rooms and

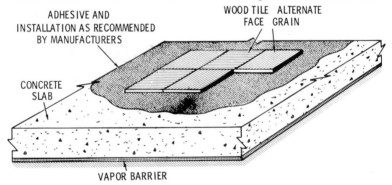

Fig. 8. Installation of wood floor tile using a vapor barrier.

wood floors are used in adjacent rooms over a subfloor of common level, a suitable base floor is required for the finish flooring. This base floor may also be tongue-and-groove flooring or plywood, and the thickness of the base floor plus the thickness of the non-wood finish floor should be equal to the thickness of the finished floors in adjacent rooms.

Linoleum

Linoluem is manufactured in thicknesses ranging from 1/16 to

1/4 inch and is generally 6 feet wide. It is made in various grades and can be purchased in plain colors, or it may be inlaid or embossed. Linoleum may be laid on wood or plywood base floor, as shown in Fig. 9, but should never be laid on a concrete slab on the ground. Since linoleum follows the contour of the base floor over which it is laid, it is essential that the base be uniform and level. When wood floors are used as a base, they should be sanded smooth and be level and dry. Plywood base floors should be carefully jointed where adjacent sheets butt together. After the base floor is correctly prepared, the linoleum is then laid, pasted, and thoroughly rolled to insure complete adhesion to the floor.

Asphalt Tile

Asphalt tile is widely used as a covering over concrete floor slabs, and is occasionally used over wood subfloors. It is the least costly of the commonly used floor covering materials. This tile is about 1/8-inch thick and 9 × 9 or 12 × 12 inches square. Most types of asphalt tile are damaged by grease and oil, and for this

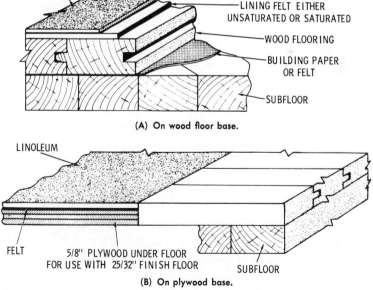

(A) On wood floor base.

(B) On plywood base.

Fig. 9. Linoleum finish on floor.

reason are not recommended for use in kitchens. In laying asphalt tile, it is important that the subfloor or base over which the tiles are to be laid is suitably prepared. Most manufacturers provide directions on the preparation of the base and furnish the type of adhesive that is most suitable to their product. The tile should be laid in accordance with the manufacturer's directions. Where a wax finish is desired, liquid self-polishing wax should be used; *never a paste wax.*

Rubber Tile

Rubber-tile flooring is resilient, noiseless, waterproof, and it stands up well under hard usage. It may be laid over concrete floor slabs, except slabs on the ground, or over wood subfloors. The finish may be plain or marblized in various designs with the colors running throughout the body of the tile. The tiles are made

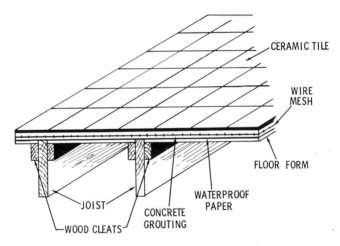

Fig. 10. Installation of ceramic tile.

in square shapes ranging in size from 4 × 4 to 18 × 18 inches and in rectangular shapes ranging from 9 × 18 to 9 × 36 inches. Their thickness is from 1/8 to 3/16 inch. The tiles are generally laid in a waterproof rubber cement and thoroughly rolled.

Rubber tile should be installed on a wood subfloor above grade. If the subfloor is plywood, rubber tile may be laid directly on the wood surface, making sure that the plywood joints do not coincide with joints in the rubber tile. If tile is laid on a solid wood subfloor, it is recommended that the floor first be sanded smooth, sealed, and then a layer of 15- to 30-pound saturated lining felt bonded to the subsurface. Tile should be installed in accordance with the manufacturer's recommendations as to both methods and materials.

CERAMIC TILE

Ceramic tile are made in different colors and with both glazed and unglazed surfaces. They are used as a covering for floors in bathrooms, entry ways, vestibules, and fireplace hearths. Ceramic tile presents a hard and impervious surface. In addition to standard sizes and plain colors, many tile are especially made to carry out architectural effects. When ceramic-tile floors are used with wood-frame construction, a concrete bed of adequate thickness must be installed to receive the finishing layer.

The minimum thickness of the concrete base and tile should not be less than 1-1/4 inches. To acquire sufficient space for the concrete bed, it is necessary to drop the wood subflooring between the joists, as shown in Fig. 10, so that the finish floor will be level with the floors in the adjoining rooms. A saturated felt should be applied over the wood subfloor. The concrete bed may be 1 part cement to 3 parts sand, or 1 part cement, 2 parts sand, and 4 parts pea gravel. A wire mesh should be used for reinforcement.

PORCH FLOORING

The material should be so installed as to resist the effects of moisture and rain. It is customary to use matched flooring strips with the joints well sealed with white lead. The slope of the floor will then permit water to drain off rapidly. In sections where heavy snows are frequent, floor strips are sometimes laid with a space of 1/8 or 1/4 inch between, to allow the melting snow to drain off more readily. Porch flooring will be either 1 inch (actually 25/32) or 1-1/4 inches (actually 1-1/16 inches) thick. It is generally better practice to use 1-1/4-inch material because of its weather-resisting qualities.

The "B and better" grade, preferably with a high percentage of heartwood, of any of the soft wood species ordinarily used for flooring is suitable for porch floors. Edge grain floor has better wearing qualities than flat-grain. Outside floors should be kept continuously painted. Splintering caused by wear on flat-grain material is less likely to occur with well-painted floors than with unpainted floors subject to hard wear. For this reason it is often desirable to use flat-grain material, but it should be heartwood for better decay-resisting qualities.

SUMMARY

After the foundation, sills, and floor joists have been constructed, the subfloor is generally laid diagonally on the floor joists. Boards 6, 8, or 12 inches wide, and generally 1 inch in thickness, are used. Plywood is also used in some cases because of its size, weight, and stability, plus the time and labor saved in application. It adds considerable strength to the floor since the weight is distributed over a wider area.

Hardwood flooring most commonly used are red and white oak, hard maple, beech, and birch. Most flooring is generally 25/32 inches thick with a face width of 2¼ inches. The finished flooring is usually installed after all of the plastering is complete, after windows and doors are in place, and after most of the interior trim has been installed.

When the flooring is ready for installation, the subfloor should be thoroughly cleaned. A layer of deadening felt or building paper will stop a certain amount of dust, and will somewhat deaden the sound, plus help to eliminate floor squeaks. Where a crawl space is used, it will increase the warmth of the floor by preventing air infiltration.

REVIEW QUESTIONS

1. What type of wood is generally used for finish flooring?
2. What is the thickness of hardwood flooring?
3. Why is building paper or deadening felt used between flooring?
4. How are sound-proof floors constructed?
5. Why is subflooring generally installed diagonally to floor joists?

CHAPTER 9

Interior Walls and Ceilings

The term *interior walls,* in this case, means *the interior covering placed over the studding or joists to form a finished surface.* This usually consists of plaster and metal lath, gypsum board, paneling and ceramic tile.

LATH

When plaster is used to form the wall surface, it is held in place by lath. There are several kinds of lath, classed according to material:

1. Expanded Metal.
2. Gypsum.

PLASTER REINFORCING

Since some drying may take place in wood framing members after the house is completed, some shrinkage can be expected. This, in turn, may cause the plaster to crack around openings and corners. To minimize this cracking, expanded metal lath is used in certain key positions over the plaster base material as reinforcement. Strips of expanded metal lath may be used over the window and door openings, as shown in Fig. 1. A strip 8″ × 20″ placed diagonally across each upper corner, and nailed lightly in place, should be effective. Fig. 2 illustrates numerous types of expanded metal lath.

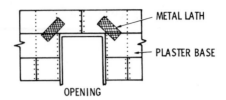

OPENING

Fig. 1. *Reinforcement of plaster over openings using expanded metal lath.*

Inside corners at the juncture of walls and ceilings should be reinforced with corners of metal lath or wire fabric, as shown in Fig. 3, except where special clip systems are used for installing the lath. The minimum width of the lath in the corners should be 5 inches, or 2 1/2 inches on each surface or internal angle, and should also be lightly nailed in place. Corner beads, as shown in Fig. 4, of expanded metal lath or perforated metal, should be installed on all exterior corners. They should be applied plumb and level. The bead acts as a leveling edge when the walls are plastered and reinforces the corner against mechanical damage.

Metal lath should be used under large flush beams, as shown in Fig. 5, and should extend well beyond the edges. Where reinforcing is required over solid wood surfaces, such as drop beams, the metal lath should either be installed on strips or else self-furring nails should be used to set the lath out from the beam. The lath should be lapped on all adjoining gypsum lath surfaces.

Expanded metal lath applied around a bath tub recess for ceramic tile application should be used over a paper backing, as shown in Fig. 6. Studs should be covered with 15-pound asphalt saturated felt applied shingle style. The scratch coat should be Portland-cement plaster, 5/8" minimum thickness, and integrally waterproofed. The scratch coat should be dry before the ceramic tile is applied. Expanded metal lath consists of sheet metal that has been slit and expanded to form innumerable openings for the keying of the plaster. It should be painted or galvanized, and its minimum weights for 16-inch stud or joist spacing are as follows:

Metal lath is usually 27" × 96" in size. Other materials, such as wood lath, woven wire fabric, and galvanized wire fabric, may

144

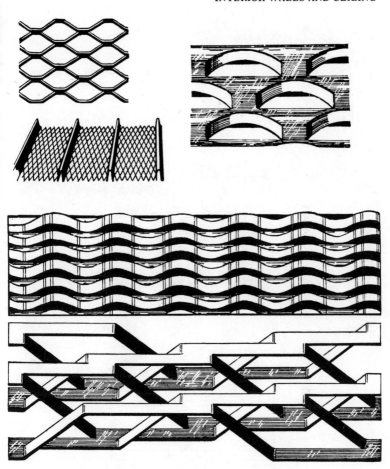

Fig. 2. Various types of expanded metal lath.

Fig. 3. Illustrating the use of cornerites made from metal lath.

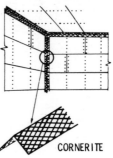

CORNERITE

145

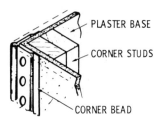

PLASTER BASE

CORNER STUDS

CORNER BEAD

Fig 4. Showing the use of a corner bead.

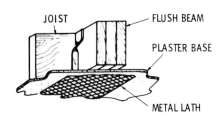

JOIST

FLUSH BEAM

PLASTER BASE

METAL LATH

Fig. 5. Illustrating the use of expanded metal lath under large flush beams.

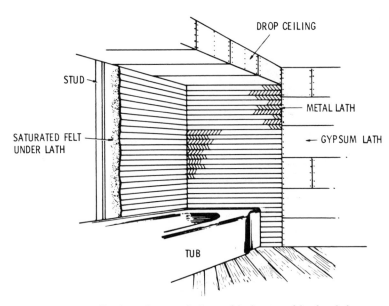

DROP CEILING

STUD

METAL LATH

SATURATED FELT UNDER LATH

GYPSUM LATH

TUB

Fig. 6. Application of expanded metal lath around bath tub for the installation of ceramic tile.

Use	Pounds per square yard
Walls	2.5
Ceilings	3.4
Ceiling with flat rib	2.75

also be used as a plaster base. Various types of woven wire lath are shown in Fig. 7.

INSTALLING GYPSUM BASE

A plaster finish requires some type of base upon which the plaster can be spread. The base must have bonding qualities so that the plaster adheres or is keyed to the base that has been fastened to the framing members. One of the popular types of plaster base which may be used on the side walls and ceilings is gypsum-board lath. Such lath is generally 16″ × 48″, and is applied horizontally (across) the frame members. This type of board has a paper face with a gypsum filler. For studs and joists with spacing of 16 inches on center , 3/8″ thickness is used, and for 24 inches on center spacing, 1/2″ thickness is used. This material can be obtained with a foil backing that serves as a vapor barrier, and if it faces an air-space, it has some insulating value. It is also available with perforations, which improves the bonding strength of the plaster base.

Insulating fiberboard lath may also be used as a plaster base. It is usually 1/2″ thick and generally comes in strips of 18″ × 48″. It often has a shiplap edge and may be used with metal clips that are located between studs or joists to stiffen the horizontal joints. Fiberboard lath has a value as insulation and may be used on the walls or ceilings adjoining exterior or unheated areas.

Gypsum lath should be applied horizontally, with the joints broken as shown in Fig. 8. Vertical joints should be made over the center of the studs or joists, and should be nailed with 13-gauge gypsum lathing nails 1 1/8″ long and having a 3/8″ flat head. Nails should be spaced 4″ on center, and should be nailed at each stud or joist crossing. Lath joints over the heads of door and window openings should not occur at the jamb lines. In-

147

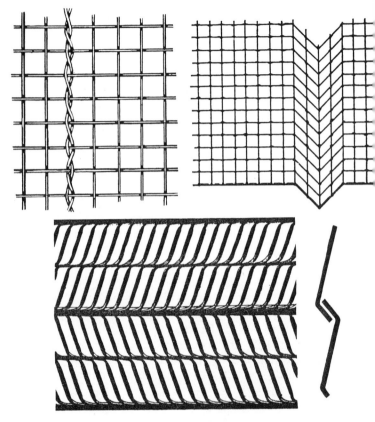

Fig. 7. Various types of woven wire lath.

sulating lath should be installed much like gypsum lath, except that 13 gauge 1 1/4″ blued nails should be used.

PLASTER GROUNDS

Plaster grounds are strips of wood the same thickness as the lath and plaster, and are attached to the framing before the plaster is applied. Plaster grounds are used around window and door openings as a plaster stop, and along the floor line for attaching the baseboard. They also serve as a leveling surface when plastering, and as a nailing base for the finish trim, as shown in Fig. 9.

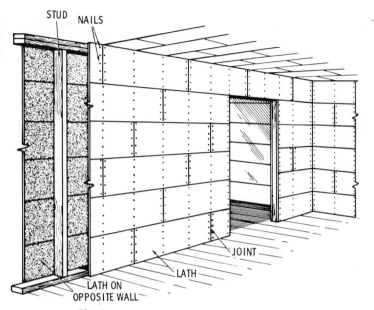

Fig. 8. *Application of gypsum plaster base.*

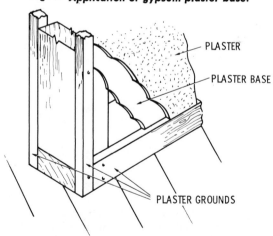

Fig. 9. *Plaster grounds used at doors and floor line.*

There are two types of plaster grounds; those that remain in place (Fig. 10), and those that are removed after plastering is completed (Fig. 11). The grounds that remain in place are usually 7/8″ thick and may vary in width from 1 inch around openings to 2 inches for those used along the floor line. These grounds are nailed securely in place before the plaster base is installed. Where a painted finish is used, finished door jams are sometimes placed in the rough openings, and the edges of the jambs serve as the grounds during the plastering operation.

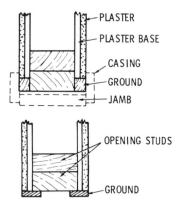

Fig. 10. Illustrating plaster grounds that stay in place and are used as a nailing base for the trim.

Fig. 11. Showing the plaster grounds that are removed after plastering

NAILING LATH

Lath nailers are horizontal or vertical members to which lath, gypsum boards, or other covering materials are nailed. These members are required at interior corners of the walls and at the juncture of the wall and ceiling. Vertical lath nailers may be composed of studs so arranged as to provide nailing surfaces, as shown in Fig. 12. This construction also provides a good tie between walls.

Another vertical nailer construction consists of a 2″ × 6″ lathing board that is nailed to the stud of the intersecting wall, as shown in Fig. 13. Lathing headers are used to back up the board. The header should be toe-nailed to the stud. Doubling of the ceiling joist over the wallplates provides a nailing surface for interior finish material, as shown in Fig. 14. Walls may be tied to the ceiling framing in this method by toe-nailing through the joists into the wall plates.

Another method of providing nailing surface at the ceiling line is similar to that used on the walls. A 1″ × 6″ lathing board

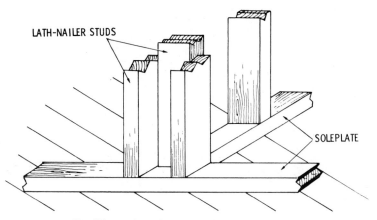

Fig. 12. Lath nailers at wall intersections.

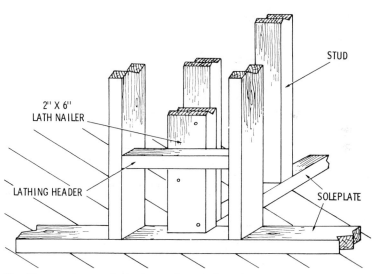

Fig. 13. Another method of installing lath nailers at intersection walls.

is nailed to the wall plate as shown in Fig 15. Headers are used to back up this board, and the header, in turn, can be tied to the wall by toe-nailing into the wall plate.

151

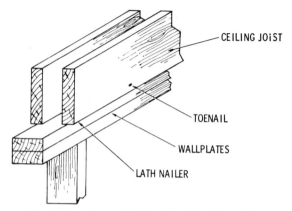

CEILING JOIST

TOENAIL

WALLPLATES

LATH NAILER

Fig. 14. Horizontal lath nailers for plaster base formed by ceiling joists.

PLASTERING MATERIAL AND METHOD OF APPLICATION

Plaster for interior finishing is made from combinations of sand, lime or prepared plaster, and water. Waterproof finish wall materials are available and should be used in bathrooms, especially in showers or tub recesses when tile is not used, and sometimes in the kitchen wainscot. Plaster should be applied in three coats or a two-coat double-up work. The minimum thick-

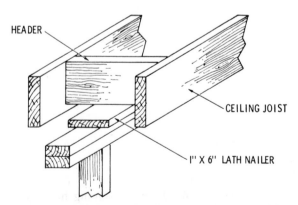

HEADER

CEILING JOIST

1" X 6" LATH NAILER

Fig. 15. Horizontal lath nailers for ceiling provided by lathing boards.

152

ness over a lath or masonry should be 1/2 inch. The first plaster coat over the metal lath is called the *scratch coat,* and is scratched after a slight set has occurred to insure a good bond for the second coat. The second coat is called the *brown or leveling coat,* and it is during the application of this coat that the leveling is done.

The double-up work, combining the scratch and brown coat is used on gypsum or insulating lath, and the leveling and plumbing of walls and ceiling are done during the application of this work. The final or finish coat consists of two general types—the sand-float and the putty finish. In the sand-float finish, lime is mixed with sand and results in a textured finish, with the texture depending on the coarseness of the sand used. Putty finish is used without sand and has a smooth finish. This is commonly used in kitchens and bathrooms, where a gloss paint or enamel finish is often employed, and in other rooms where a smooth finish is desired. The plastering operation should not be done in freezing weather without the use of constant heat for protection from freezing. In normal construction, the heating unit is in place before plastering is started.

Insulating plaster, consisting of a vermiculite, perlite, or other aggregate used with the plaster mix, may also be used for wall and ceiling finishes. This aggregate properly mixed with the plaster produces small hollow air pockets which act as insulating material. The vermiculite is a material developed from a mica base which is exploded in size, and when mixed with the plaster, reduces the added weight in a conventional plastered wall or ceiling.

The following points in plaster maintenance are worthy of attention:

1. In a newly constructed house, a few small plaster cracks may develop during or after the first heating season. These cracks are usually due to drying and shrinking of the structural members. For this reason, it is advisable to wait until after a heating season before painting the plaster. These cracks can then be filled before painting has begun.
2. Because of the curing period ordinarily required for a plastered wall, it is not advisable to apply oil-base paints until

at least 60 days after plastering is completed. Water-mix, or resin-base paints may be applied without the necessity of an aging period.

3. Large plaster cracks often indicate a structural weakness in the framing. One of the common areas that may need correction is around a basement stairs. Framing may not be adequate for the loads of the walls and ceilings. In such cases, the use of an additional post and pedestal may be required to correct this fault. Inadequate framing around fireplaces and chimney openings, and joists that are not doubled under partitions, are other common sources of weakness.

DRY-WALL FINISH

Dry-wall finish is a material that requires little if any water for the application. More specifically, dry-wall finish may be gypsum board, plywood, fiberboard or wood in various sizes and forms. The use of thin sheet materials such as plywood or gypsum board requires that the studs and ceiling joists be in alignment. Good solid rigid sheathing on the exterior walls will accomplish this on the studs, and strongback bracing on the ceiling joists (Fig. 16), will keep the ceiling joists level and in alignment.

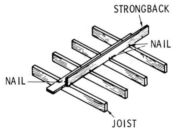

Fig. 16. Strongback bracing for ceiling joists.

Minimum thicknesses for materials often used are as follows:

Finish	Material thickness when framing is spaced		
	16 inches	20 inches	24 inches
Plywood	1/4″	1/4″	3/8″
Gypsum	3/8″	1/2″	1/2″
Fiberboard	1/2″	3/4″	3/4″

Gypsum board is a sheet material composed of a gypsum filler faced with paper. Sheets are usually 4 feet wide and they can be obtained in lengths from 8 to 12 feet long. The edges along the length of the sheet are recessed to receive the joint cement and tape. Although gypsum board may be used in 3/8" thickness, a 1/2" board offers greater resistance to deflection between framing members. Gypsum board may be applied to the walls either vertically or horizontally. Vertical applications with 4-foot wide sheets will cover three studs when they are spaced 16 inches on center. The edges should be centered on the studs, and only moderate contact should be made between the edges of the sheet. Fivepenny cement-coated nails (1 5/8" long) should be used with 1/2" gypsum board, and fourpenny (1 3/8" long) with 3/8" board. Nails should be spaced 6 to 8 inches for side walls, and 5 to 7 inches for ceiling application, as shown in Fig. 17.

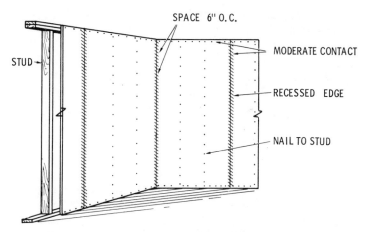

Fig. 17. *Application of vertical applied dry-wall finish.*

The horizontal method of application is best adapted to rooms in which full-length sheets can be used, to minimize the number of vertical joints. When joints are necessary, they should be made at windows and doors. Nail spacing is the same as that used in vertical application. Where framing members are more than 16 inches on center, solid blocking should be used along the horizontal joint, as shown in Fig. 18.

Another method of gypsum board application includes an under course of 3/8″ material applied vertically by nailing. The finish 3/8″ sheet is applied horizontally, usually in room-size lengths by means of an adhesive, with only enough nails used to hold the sheet in place until the adhesive is dry. The manufacturer's recommendations should be followed. In the finishing operation, the nails should be set about 1/16″ below the face of the board. This set can be obtained by using a slightly crowned hammer, as shown in Fig. 19. Setting the nail in this manner, a slight dimple

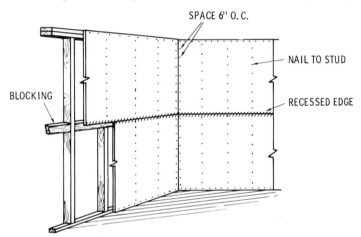

Fig. 18. *Application of horizontal applied dry-wall finish.*

Fig. 19. *Nail set with crowned hammer on dry wall.*

is formed in the face of the board without breaking the paper surface. The setting of the nail is particularly important for center nails, since the edge nailing will be covered with tape and joint cement.

Joint cement comes in powder form and is mixed with water to a soft putty consistency. The procedure for taping joints (Fig. 20) is as follows:

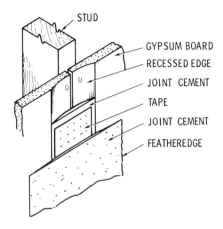

Fig. 20. Applying tape and cement to wall joints.

1. Use a wide putty knife (4″ to 6″), and spread the cement in the recess starting at the top of the wall.
2. Press the tape into the recess with the putty knife until the joint cement is forced through the perforations.
3. Cover the tape with additional cement, feathering the outer edge.
4. Allow the cement to dry and apply a second coat and feather the edges. A steel trowel is sometimes used. For best results, a third coat may be applied, feathering beyond the second coat.
5. After the joint cement has dried, sand it smooth.
6. For hiding nail indentations in the center of the board, fill with joint cement and sand smooth when dry.

Interior corners may be treated with tape. Fold the tape down the center to a right angle, as shown in Fig. 21. Apply cement to

Fig. 21. Folding joint tape for use on interior corners.

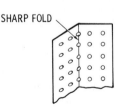

both sides and press the perforated tape into the cement. Smooth the tape down with a trowel. After the cement has dried, apply a second coat. Sand the cement when dry, and apply a third coat if needed. The interior corners between the wall and ceiling may be concealed by the use of a moulding. When moulding is used as shown in Fig. 22, it is not necessary to tape the joints. Metal corner beads should be used on exterior corners and also for openings where a casing or trim is not used.

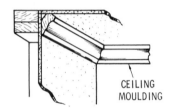

CEILING
MOULDING

Fig. 22. *Illustrating the use of moulding at the wall and ceiling corner when using dry-wall construction.*

Where time and expense will not warrant taping and cementing dry-wall joints, they can be covered with a batten or moulding strip. Fig. 23 shows various types of mouldings that can be used to cover the joints. Fig. 23 A illustrates the common batten strip, which is much easier to install because of the butt joints. The moulding strips should have mitered joints, but if expense permits, should be coped to insure a perfect fit. Fig. 24 illustrates the effect of treating the joints by nailing moulding strips to all the joints. It is possible to vary the amount of paneling strips and employ any design that is desired.

MASONRY WALLS AS PLASTER BASE

Concrete masonry provides an excellent base for plaster. The surface of any masonry unit should be rough to provide a good mechanical key, and should be free from paint, oil, dust and dirt, or any other material that might prevent a satisfactory bond. Proper application of plaster requires:

1. That the plaster base material bonds and becomes an integral part of the base to which it is applied, as shown in Fig. 25.
2. That it be used as a thin reinforced base for the finished surface.

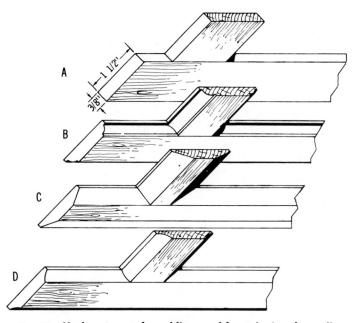

Fig. 23. Various types of moulding used for stripping dry-wall.

Old masonry walls which have been softened by weathering, or surfaces that cannot be cleaned thoroughly, must be covered with metal reinforcement before applying the plaster. Metal reinforcement should be applied to wood furring strips, as shown in Fig. 26. The metal reinforcement should be well braced and rigid to prevent cracking the plaster. This type of construction will give some insulating qualities due to the air space between masonry and plaster.

PLYWOOD PANELS

Plywood may be used in large sheets and applied horizontally or vertically. With horizontal application, the joints should be backed with solid blocking. Joints may have a V-edge, or a moulding may be used to cover the joints. Plywood should be nailed to the framing members with 1 1/4" brads for 1/4" plywood, and the nails should be set. Grooves may be cut with a table saw or a power hand saw to vary the pattern.

159

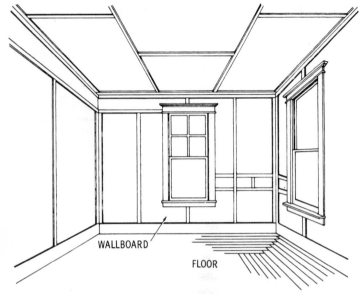

WALLBOARD

FLOOR

Fig. 24. A room illustrating the dry-wall panel joints stripped with a batten strip or moulding.

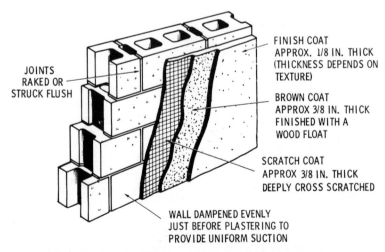

JOINTS
RAKED OR
STRUCK FLUSH

FINISH COAT
APPROX. 1/8 IN. THICK
(THICKNESS DEPENDS ON
TEXTURE)

BROWN COAT
APPROX 3/8 IN. THICK
FINISHED WITH A
WOOD FLOAT

SCRATCH COAT
APPROX 3/8 IN. THICK
DEEPLY CROSS SCRATCHED

WALL DAMPENED EVENLY
JUST BEFORE PLASTERING TO
PROVIDE UNIFORM SUCTION

Fig. 25. The application of plaster to concrete masonry.

WOOD AND FIBERBOARD

The so-called plank application may consist of fiberboard or

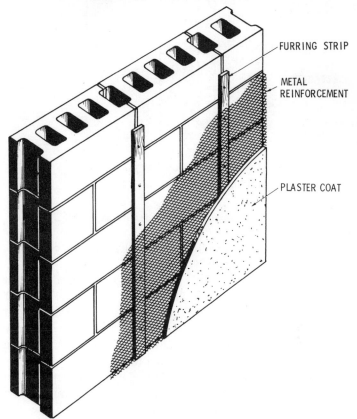

FURRING STRIP

METAL REINFORCEMENT

PLASTER COAT

Fig. 26. Furring strips applied to concrete masonry for the application of plaster.

wood of various specifications, as shown in Fig. 27. It may be used vertically or horizontally, and the application of both wood or fiberboard is facilitated by a tongue-and-groove system, often with additional edge moulding. When vertical application is used, nailers should be provided between the studs and the thickness of the material should be in accordance with the vertical span be-

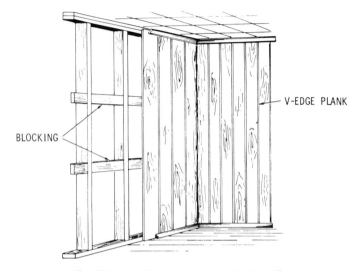

V-EDGE PLANK

BLOCKING

Fig. 27.   Application of wood plank walls.

tween the nailers. Blind nailing is generally used to conceal the nail heads.

Fiberboard supplied in squares or rectangles is called *tileboard*. Sizes are 12″ × 12″, 12″ × 24″, 16″ × 16″ and 16″ × 32″. The tile is often tongue-and-grooved and is supported by concealed nails, clips, or staples. Fiberboard is sometimes used on ceilings or above a wainscot on walls. Fiberboard for exposed interior surfacing may be obtained with a factory painted finish, or it may be repainted on the job for the desired interior effects. Acoustical fiberboard tile is sometimes used on ceilings. Various types and patterns of wood are available for application on the walls to obtain the desired decorative treatment. For informal effects, knotty pine, white pocket fir, wormy chestnut, and cypress are often used to cover one or more sides of a room. This type of paneling can be finished natural or sometimes stained and varnished. Interesting wall treatment may be obtained, as shown in Fig. 28, by cutting in patterns or by changing the direction of the grain. This particular treatment is often effective on a fireplace wall or in a den, but it should not be overdone because its effectiveness will be lost when all walls are treated too much alike.

PLYWOOD
SQUARES

STUDS

Fig. 28. Application of plywood squares.

SUMMARY

Metal lath is generally used around openings and corners to minimize cracking the plaster. Since wood draws moisture from the plaster, metal lath is used to reinforce the plaster. This type of material is also used to cover wide cracks and open joints in lath.

Gypsum board is another type of lath, which is usually 16″ × 48″, and is applied horizontally to the wall studs. It is obtained in ⅜″ and ½″ thickness, depending on wall stud spacing. Gypsumboard is also used as a dry-wall finish. It is obtained in 4 × 8 or 4 × 12 foot sheets. The edges along the length of the sheet are recessed to receive the joint tape and cement.

REVIEW QUESTIONS

1. What are plaster grounds, and why are they used?
2. Why is gypsumboard or plaster board used instead of wood lath?
3. What is dry-wall and why is it used?
4. Why is metal lath used and where is it installed?
5. What is joint tape and how is it used?

163

CHAPTER 10

Circular Saws

The circular saw is one of the most popular machines in any woodworking shop or plant. Plants of any size usually have one or more power saws used exclusively for ripping and one or more for crosscutting. Beveling and mitering can be done with circular saws of the tilting-arbor type, while grooving and dadoing can also be done by means of special cutters.

Circular saws are made in a large variety of types and sizes, among which are the universal and variety saw. The universal type is equipped with two arbors to permit mounting of both the ripsaw and the crosscut blades, either of which is brought into use by simply turning a handwheel, which controls the position of the arbor. The variety saw employs a single arbor and can be fitted with mortising and boring attachments.

The main structural features of either machine are similar, however, and include the arbor, saw blades, table, ripping fence, crosscutting and mitering fence (or gauge), and a substantial base.

In recent years, circular saws mounted on a radial arm above the work have come into wide use. These are usually equipped for variable angle cutting and can also be used for dadoing or other cutting operations with special attachments.

CONSTRUCTION

The circular saw, Fig. 1, consists generally of a cast-iron base or frame on which a table is mounted and an arbor or shaft which carries the saw blade or other cutter. The arbor or shaft revolves in two bearings, which are bolted to the frame. It is driven by a

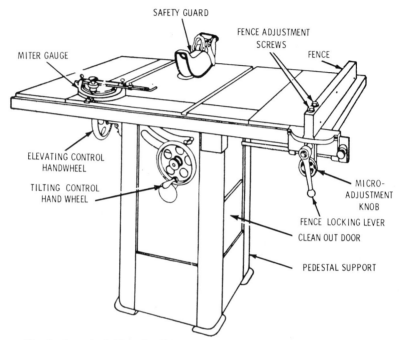

Fig. 1. A typical 10-inch, tilting-arbor saw. A circular saw of this type has a capacity of cuts 3-1/8 inches deep; 2-1/8 inches at a 45° angle. Dado head cuts 1-1/8 inches deep. The saw is driven by means of a 3450-rpm, self-contained electric motor, which transfers power to the saw by means of a triple V-belt. The speed of the saw is 3800 rpm. It is equipped with a cast-iron table, having both front and side extensions, in addition to a safety guard, fence, and miter gauge. The entire arbor and drive unit may be tilted to any desired angle for angle cutting by means of an easily accessible handwheel.

belt which passes over a pulley; this pulley is fastened to the shaft between two bearings.

In the conventional design, the saw table can be tilted to an angle of 45°, whereas on the tilting-arbor type, the table is fixed in a stationary, level position; tilt positions are obtained by tilting the saw blade. This later design is not only of great convenience to the operator, but also supplies a safety factor, since he does not have to work in an awkward position.

When wood is cut on a circular saw, it must be firmly held against a metal guide or fence, which can be set at any convenient distance from the saw blade. When the fence is used as a guide for cutting boards lengthwise, the operation is known as *ripping*. Most saw tables are equipped with two slots or grooves to accommodate a miter gauge, which is used as a guide when sawing across a board; this operation is known as *crosscutting*.

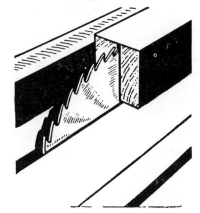

Fig. 2. Ripping operation on square stock.

RIPPING OPERATION

One of the most useful operations of the circular saw is that of ripping stock to its required width, as shown in Figs. 2 to 4. This operation is generally accomplished as follows:

1. The fence is set to the graduated scale at the front of the table to cut the required width.
2. The saw is adjusted (raised or lowered) to project approximately ¼ inch above the stock to be ripped.
3. The splitter and saw guard are positioned to insure safe operation.
4. The operation is started by holding the work close to the fence and pushing it toward the rotating saw blade with a firm, even motion. A smooth uniform speed of feed should be used; avoid jerky movements and jamming the work through too quickly. The operator should not stand directly behind the saw blade but should take a position slightly to either side and hold the stock near its end, so that one hand will pass to the right and the other hand will pass to the left of the saw blade.

CROSSCUTTING OPERATION

Square crosscutting work on the circular saw is performed by placing the work against the miter gauge and then advancing both the gauge and the work toward the rotating saw blade. The gauge may be used in either table groove, although most operators prefer the left-hand groove for average work. It is highly essential that the miter gauge is set correctly in order to obtain a square cut; therefore, it is customary to test the work by means of a try square before proceeding.

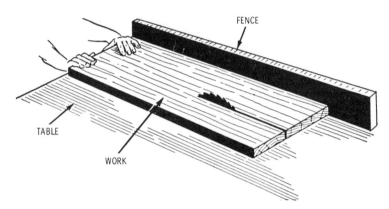

Fig. 3. Ripping operation on wide stock.

MITERING OPERATIONS

Most miters are cut to an angle of 45°, because four pieces cut at this angle will make a square or a rectangle when assembled. Miters are cut by setting the miter gauge at the required number of degrees. The angle of the miter for any regular polygon is obtained by dividing 180° by the number of sides and subtracting the quotient from 90°. For example, to find the angle for the miter of a pentagon (5 sides), we have $90 - (180/5) = 54°$. Similarly, the angle for the miter of an octagon (8 sides) will be $90 - (180/8) = 67.5°$. The miter gauge may be used in either of the table grooves and also may be set on either side of the center position.

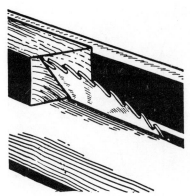

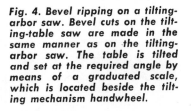

Fig. 4. Bevel ripping on a tilting-arbor saw. Bevel cuts on the tilting-table saw are made in the same manner as on the tilting-arbor saw. The table is tilted and set at the required angle by means of a graduated scale, which is located beside the tilting mechanism handwheel.

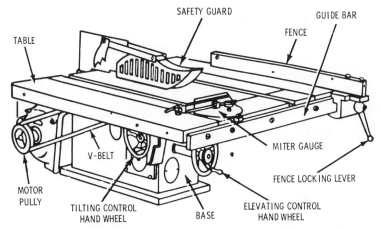

Fig. 5. A typical 8-inch, tilting-table saw. It can achieve a speed of 3450 rpm and is equipped with a cast-iron table, fence, miter gauge, and a safety guard, which includes two antikickback pawls and a splitter. The elevating mechanism consists of a handwheel which tilts the table by a screw-and-nut action.

When great accuracy is required on miter cuts, special miter-clamp attachments are used.

GROOVING OPERATIONS

These operations consist of making grooves that are wider than those cut by ordinary saw blades. Grooves of varying widths are

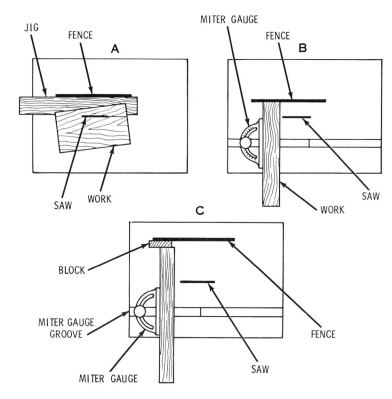

Fig. 6. Typical sawing operations on the circular saw. In taper ripping (A), work that is to be ripped on a taper cannot be guided against the fence but must be held in a tapering jig. The same idea can be applied to a number of other forms. When cutting shoulders, the stock should first be squared on one end and cut to the desired length. The miter gauge is used in conjunction with the fence to bring the shoulder cut the correct distance from the end (B). When crosscutting wide pieces to length, one end should first be squared and the gauge adjusted to the required length. Narrow pieces may be sawed to length by placing a block against the fence (C); the distance from the saw blade to the block is then the required length.

commonly cut on a circular saw by employing a special attachment known as a *dado head*. This "head" is made up of two outside cutters and three or four inside cutters, as shown in Fig. 7, by means of which grooves varying in width of from $\frac{1}{8}$ to $\frac{13}{16}$ inch or larger can be made by using different combinations of cutters. The outside cutters generally have eight sections of cutting teeth

Fig. 7. A typical dado-head assembly. A dado head, as shown, is made up of two outside blades 1/8 inch thick together with one or more cutting fillers 1/16 inch and 1/8 inch thick, depending on the width of the groove desired. Grooves varying in widths up to 1 inch can be cut.

and four raker teeth. The eight sections of cutting teeth are ground alternately left and right to divide the cut. Inside cutters of ⅛ inch and ¼ inch are usually swage set for clearance.

Grooves which are cut across the grain are termed *dadoes*. They are usually cut at right angles, but may also be cut at other angles. Dadoes that do not extend entirely from one side of a board to the other side are called *stopped dadoes, blind dadoes,* or *gains*. A gain may have one open end, or both ends may be closed. Two grooving operations are shown in Figs. 8 and 9.

POWER AND SPEED

The required power for circular saws may vary widely, depending on their use. The actual power required depends on: thickness of cut, rate of feed, kind of wood, and condition of saw blade. Thus, for example, a heavy-duty ripsaw of the sawmill type will require a motor rated at 50 to 100 horsepower, whereas a typical

171

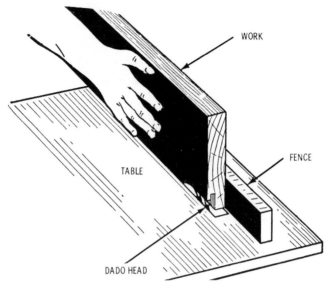

Fig. 8. The method of cutting a rabbeted joint with a dado head. Joints of this type are used extensively in drawer construction.

Fig. 9. The method of cutting wide grooves with a dado head.

8-inch circular saw used in the average home workshop will require a motor of from ⅓ to ½ horsepower. A typical spur-feed ripsaw used in a woodworking plant requires a motor of from 5 to 10 horsepower.

172

Speed is indicated by the number of revolutions per minute made by the saw and also by the number of feet traveled by the rim per minute. Motor-driven circular saws with the motor mounted directly on the spindle usually run at 3600 revolutions per minute (rpm) on 60-cycle AC current. The speed can be adjusted to a suitable value by means of two pulleys of unequal diameter—one on the motor shaft and the other on the saw arbor. It is a mistake to run any saw faster or slower than the manufacturer recommends, since saw blades are tensioned to run at a certain speed, and they produce the best results at this speed.

Rules for Calculating the Speed of Saws and Pulleys

Problem—The diameter of the driver is given; find its number of revolutions per minute.

Rule—Multiply the diameter of the driven pulley by its number of revolutions per minute, and divide the product by the diameter of the driver; the quotient will be the number of revolutions per minute of the driver.

Problem—The diameter and revolutions per minute of the driver are given; find the diameter of the driven pulley that will make the required number of revolutions per minute in the same time.

Rule—Multiply the diameter of the driver by its number of revolutions per minutes, and divide the product by the revolutions per minute of the driven pulley; the quotient will be its diameter.

Problem—Ascertain the size (diameter) of the driver.

Rule—Multiply the diameter of the driven pulley by the number of revolutions per minute desired, and divide the product by the revolutions per minute of the driver; the quotient will be the diameter of the driver.

The values given in Table 1 apply to solid-tooth saw blades.

Table 1.   Speed of Circular Saws

Size of Saw Blade	Rev. per min.
8 in.	4500
10 in.	3600
12 in.	3000
14 in.	2585

173

Rim Speed

A typical speed of 9000 feet per minute for the rim of a circular-saw blade may be laid down as a rule. For example, a saw blade that is 12 inches in diameter is 3 feet around the rim and should rotate at 3000 rpm. Of course, it should be understood that the rim of the saw blade will run a little faster than this calculation, because the circumference is slightly more than three times as large as the diameter (3.14 times the diameter, to be exact).

SUMMARY

The efficiency of a circular saw depends mainly on the type and condition of the blade used. There are generally three types of circular-saw blades; crosscut, ripsaw, and combination.

Many operations can be performed on a circular saw, such as ripping, rabbeting, dado cutting, miter cutting, molding cutting, and cutting tenons. In the conventional design, the saw can be tilted to an angle of 45 degrees, which is a safety factor since the operator does not have to work in an awkward position. When wood is cut on a circular saw, it must be firmly held against a metal guide or fence.

REVIEW QUESTIONS

1. Why is it better to tilt the saw blade than to tilt the table in miter cuts?
2. Explain the purpose of the fence.
3. What is a dado head?
4. What is rabbeting?
5. Explain the use of a feather board.

CHAPTER 11

Band Saws

Band saws are manufactured for a large variety of uses. For ripping and resawing, large machines with blades from 3 to 5 inches wide are generally used. The type most adaptable to the general wood shop is the band scroll saw, which has blades from ½ to 1 inch wide and is used particularly for cutting curved outlines and lines that are not parallel to an edge of the piece being cut.

Essentially, a band saw consists of a table, wheels, guides, saw blade, and suitable guards. On most of the larger machines, the table can be tilted, usually 45° in one direction and 10° in the other. Some, however, are built to permit bevel cutting by varying the blade angle. The band saw is generally used for resawing because of its thinner kerf.

CONSTRUCTION

A band saw, Fig. 1, such as is used in woodworking shops consists generally of an endless band of steel with saw teeth on one edge, passing over two vertical wheels and through a slot in a table. The blade is held in position for the thrust of the wood against the teeth by two guides.

The two vertical wheels over which the blade is fitted are usually made of cast iron; their rims are equipped with rubber tires and are provided with adjustments for centering the saw on the rims and for giving the saw blade the proper tension under all loads.

The table supporting the work is fastened to a casting directly above the lower wheel. It is slotted for the saw blade from the

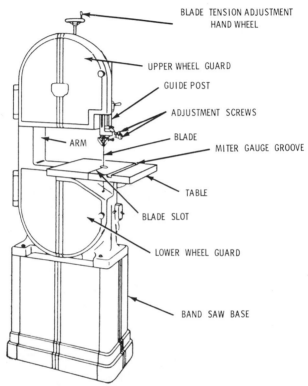

BLADE TENSION ADJUSTMENT HAND WHEEL

UPPER WHEEL GUARD

GUIDE POST

ADJUSTMENT SCREWS

BLADE

MITER GAUGE GROOVE

ARM

TABLE

BLADE SLOT

LOWER WHEEL GUARD

BAND SAW BASE

Fig. 1. A typical 14-inch band saw suitable for the small- and medium-sized woodworking shop. Tension and tilt are regulated by a convenient handwheel and a knob. The blade guides are designed to provide maximum support for the work, with the blade fully shielded for safety. The wheels are dynamically balanced, and the upper wheel unit rides on two heavy ground-steel rods with a spring cushion to absorb blade shock. The blade speed of the saw is 2535 surface feet per minute with a 1750-rpm motor.

center to one edge. To prevent the blade from twisting sideways in the slot and to give it support when cutting, the band saw is provided with guides, the design of which varies for different types of saws. Some band-saw tables are equipped with a ripping fence and some are also provided with a groove for a miter gauge.

The size of the band saw depends on the diameter of the wheels, which may vary in size from 10 to 40 inches approximately. Thus, a saw with 10-inch diameter wheels would be called a 10-inch saw. Of course, the larger the wheels, the larger in proportion are all the other parts of the saw, and consequently larger size stock can be sawed. Other important dimensions of the band saw are the table size and the height between the table and the upper blade guide.

STRAIGHT-CUTTING OPERATION

Although a band saw is essential for curved cutting, it may also be used for making straight cuts for both crosscutting and ripping, where a circular saw is not available. For all straight cuts, it is advisable to use the widest possible blade, because it is easier to follow a straight line with a wide blade. The use of the miter gauge for crosscutting wide stock, as shown in Fig. 2, follows the same general procedure as that used for similar work on the circular saw. In the absence of a miter gauge, a wide board with square ends and sides may be used, as shown in Fig. 3.

The use of an auxiliary wood fence fastened to the gauge will facilitate the handling of large boards and will result in more accurate work. Unlike the auxiliary ripping fence for the circular saw, the wood fence for the band saw should be kept low, so that it will work under the guides.

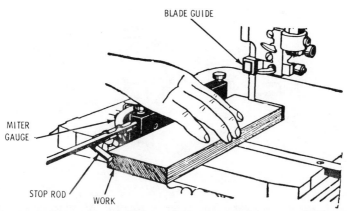

Fig. 2. Cutting to length with a miter gauge and a stop rod. The stop rod should be carried on the outer end of the miter gauge.

177

Ripping and resawing may be performed on a band saw by the use of the ripping fence furnished with most saws. When the stock

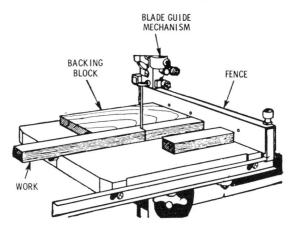

Fig. 3. The method of cutting short pieces to length using a square board or a ripping fence, when the miter gauge is not available.

is worked flat on the table, the operation is known as ripping, and when the board is worked on edge, the operation is usually known as resawing.

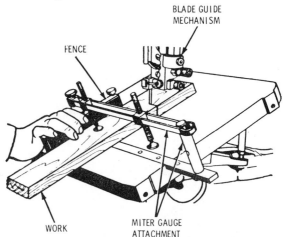

Fig. 4. Use of the miter gauge clamp attachment. An attachment of this type is useful in many crosscutting operations and particularly when cutting at an angle with the table tilted.

Another type of guide frequently used in these operations is known as the pivot block. The pivot block consists of a specially formed wood block that is clamped to the table to hold it in the desired position. The guide is set opposite the blade and at the proper distance from it to cut the required thickness, as shown in Fig. 4.

CUTTING CIRCULAR ARCS AND SEGMENTS

When cutting circular arcs, the usual procedure is to make an outline by means of a compass or divider after determining the correct radius or diameter. If several pieces all having the same curvature are to be sawed, a jig may preferably be made, in which case the circular arcs can be accurately cut without first marking the outline.

MULTIPLE-SAWING OPERATION

To secure maximum output on band-sawed work, it is necessary to use an up-to-date machine that is securely mounted on a substantial foundation to eliminate vibration. The blades are the next important part of the equipment; they must be kept in prime condition and must be properly adjusted on the wheels. When a considerable amount of wood is to be processed, an extension to the saw table is a real convenience, if not an actual necessity.

The numerous furniture parts that are frequently sawed in multiple include ornate chair and table stretchers, chair bannisters, radio-cabinet grilles, small brackets, etc. The grilles and similar items, however, are usually scrolled out on a jigsaw. Dependable machines are usually provided with accurate tension devices which assist the operator in securing volume production as well as turning out high-class work.

Some present-day jigsaws are constructed with the table and saw guides set at such an angle with reference to the machine column that extremely long material may be sawed easily.

When cutting certain classes of thick material, the best results may be obtained by sawing one piece at a time, particularly where there are pitch spots and checks and knots to be dodged. On the other hand, anywhere from 8 to 18 pieces of veneer or thin plywood can often be scrolled out simultaneously, depending on the thickness of the stock. Some favor the idea of tacking pieces together lightly before sawing, but a more satisfactory system is to cut two or more ¾-inch slits in their edges after stacking them up evenly on the saw table. A hardwood wedge driven into each of these slits will serve to hold the stock together while being processed. These two methods are illustrated in Fig. 5. It is not unusual to find workmen scrolling eight or more pieces of ½-inch plywood at one time without fastening the plies together in any manner. This procedure calls for considerable dexterity, and it is really not advisable to attempt it on work where great accuracy is essential unless the workman is an expert.

POINTERS ON BAND-SAW OPERATION

In order to obtain the maximum quantity as well as the best possible quality from the band saw, it is necessary that the operator understand its operation and be able to adjust it properly. Prior to the actual sawing, it is necessary that the operator be carefully instructed in the various safety features with which the

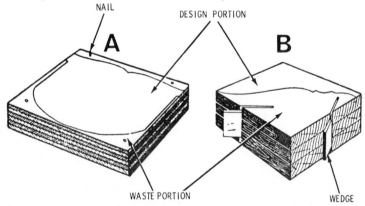

Fig. 5. *The methods of assembly commonly used in multiple-sawing operations; A, nails are driven into the waste portions of the design to hold the parts together while being sawed; B, the stock is wedged together while being sawed.*

band saw is equipped. He should be familiar with the removal of the wheel guards and, before operating the band saw, should make certain that they are securely fastened. Some band saws are equipped with a braking device whereby the drive wheel may be stopped quickly for blade changes. One type is provided with an automatic brake which instantly stops the wheels if the blade breaks.

Emphasis should be placed on the fact that the widest blade possible should be used, giving consideration to the minimum radius to be cut on a particular class of work. A rule of thumb used by many is that the width of the blade should be one-eighth the minimum radius to be cut. Therefore, if the piece on hand has a 4-inch radius, the operator would select a ½-inch blade. This rule should not be construed to mean that the minimum radius that can be cut is eight times the width of the blade, but rather that such a ratio indicates the practical limit for high-speed band-saw work.

Probably the most frequent cause of difficulty in obtaining good results with the band saw is the misalignment of the blade guides. There are several styles of guides available, each of which has certain advantages. Most guides have a hardened steel disc mounted on a ball bearing to serve as a support for the back of the blade. The purpose of the back guide is to hold the blade as the work is being cut and to prevent it from being pushed from the wheels.

When the saw blade is properly aligned on the wheels, the back guide will not be in contact with the blade. If it is in constant contact with the back support, the resulting friction will, in time, cause the back edge of the blade to become case-hardened. Such a blade with unequal tension throughout its width is susceptible to fractures and breakage. A blade in contact with the back-guide wheel when it is not cutting also indicates that there is excessive tilt in the upper wheel, which will cause too much bearing wear.

In addition to the back-support wheel, the band saw is equipped with two side pressure guides which prevent the blade from twisting as the work is cut. These are either square pieces of hardened tool steel or rollers, depending on the type of guide.

The square tool-steel type is most common and should be set so that the teeth are slightly forward from the guide pins; this is necessary so that the blade will not be dulled and the set will not be. removed from the teeth by contact with the guides. The proper clearance between the side guides and the blade can be measured by placing a piece of paper between them when adjustments are made. When the guides are properly set, the blade will not touch them while the saw is running under a no-load condition. This procedure requires perfect alignment between the upper and lower guide sets. When correctly adjusted, there should be a $\frac{1}{64}$-inch clearance between the blade and all guides.

Another type of band-saw guide that has come into limited use during the past few years is pivoted on a yoke arrangement, so that it allows the blade to twist when following curves. Such an arrangement allows more rapid and accurate cutting of curved sections with less effort than that required where the conventional type of guide is used.

The problem of obtaining correct blade tension is difficult for inexperienced operators on band saws that are not equipped with a tension gauge. Several band saws have tension scales which are calibrated to show correct tension for each of the various saw blades. If the saw is not so equipped, the operator must learn to adjust the blade tension by "feel." One method of testing blades that are $\frac{1}{2}$ inch and narrower is to raise the upper guide, place the first and fourth fingers on one side of the blade, and push the blade with the thumb on the opposite side. If the blade can be deflected slightly, the tension is about right, but if the blade cannot be flexed, it is tensioned too tightly. On blades wider than $\frac{1}{2}$ inch (with the upper guide raised 12 inches above the lower guide), the blade should flex approximately $\frac{1}{8}$ inch for proper tensioning.

Where the band saw is operated continuously, the tension may have to be increased gradually during the day, because the heat of the blade will cause it to expand and stretch. As the blade expands, tension will decrease and cutting will become more dfficult. At the end of the day's work, the blade tension should be relieved, since the blade will contract as it cools and may fracture if the tension is too great.

Since smooth band-saw operation depends on the proper adjustment of the guides and the proper tensioning and tracking of the blade, the operator who does not understand how to adjust the band saw cannot be expected to turn out first-class work.

The operator should also be able to select the correct blade for the class of work at hand. As previously mentioned, the widest blade possible should be used, taking into consideration the radius of the curves to be cut. The reason for this is that on straight or gently curving portions of work, it is much easier to follow the contour with a wide blade, since narrow blades often have a tendency to wander.

Band-saw blades are commonly classified as 4-, 5-, 6-, or 7-tooth blades. This designation refers to the number of teeth per inch of blade length, as shown in Fig. 6. Where smooth cuts are desired, the 6- or 7-tooth variety should be used. Where speed is

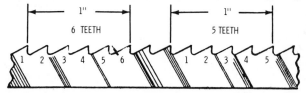

Fig. 6. The method of designating the number of teeth per inch.

of more importance than the smoothness of a cut, a 4-tooth blade should be employed, because its larger teeth will cut more rapidly.

SUMMARY

Band saws are used for cutting curved outlines and lines not parallel to a straight edge. The type most adaptable to the general wood shop use is called a band scroll saw. They generally use a blade ½ to 1 inch in width and 10 to 40 inches in diameter, depending on the diameter of the upper and lower drive wheel.

Although a band saw is essential for curved cutting, it may also be used for making straight cuts for both crosscutting and ripping when a circular saw is not available. When using a band saw for straight cutting, it is easier to follow a straight line with a wide blade.

REVIEW QUESTIONS

1. What is the big advantage in using a band saw?
2. How many teeth per inch on the common band-saw blade?

3. How should tension on the saw blade be checked?
4. How is straight cutting accomplished?
5. How is miter cutting accomplished?

CHAPTER 12

Jigsaws

The jigsaw differs radically in construction from the band saw, although the type of work for which it is designed is quite similar. The jigsaw is more adaptable than the band saw for cutting small, sharp curves because much smaller and finer blades may be used. Inside cutting is also better accomplished on the jigsaw, since the blade is easily removed and inserted through the entrance hole bored in the stock.

CONSTRUCTION

The jigsaw, as illustrated in Fig. 1, consists essentially of a base or frame, a driving mechanism, a table, a tension mechanism, guides, and a saw blade. A typical table-model jigsaw is shown in Fig. 2. The driving mechanism, as shown in Fig. 3, has a motor-driven wheel that is connected by a steel rod to a bar, called the "cross-head," which moves up and down between two vertical slides. This arrangement converts the rotating motion of the motor into a reciprocating (up-and-down) movement of the blade and provides an efficient cutting action.

The table built around the blade is usually designed for tilting at angles up to 45°. The size of the jigsaw is generally expressed in terms of the throat opening, that is, the distance from the blade to the edge of the supporting arm. The distinguishing feature of saws of this type, as previously noted, is that the blade moves with a reciprocating motion instead of moving continuously in one direction, as in the case of circular and band saws. Accordingly,

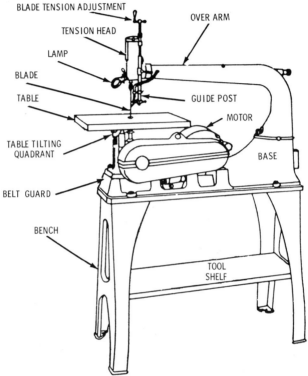

Fig. 1. The component parts of a typical jigsaw. A jigsaw of this type uses a standard 6-inch blade, but it can accommodate any blade from 5 to 15 inches because of hollow vises and shafts. The cast-iron table can be tilted 45° either way and is equipped with a graduated quadrant showing the exact number of degrees of tilt. The upper head adjusts on dovetail ways by means of a hand crank and a knurled locknut. Blade tension is shown on a scale and may be regulated for minimum vibration while the machine is running. The drive mechanism is of the reciprocating type with link and counterbalanced crank. The air pump is powered by the main drive shaft and keeps the cutting line clear at all saw-blade speeds. The guides are completely adjustable for front and side sawing and are equipped with a soft refaceable blade support and a ring-type hold-down device.

only one-half of the distance traveled by the saw blade is effective in cutting.

With the jigsaw and its numerous attachments, there are comparatively few cutting operations that cannot readily be accom-

186

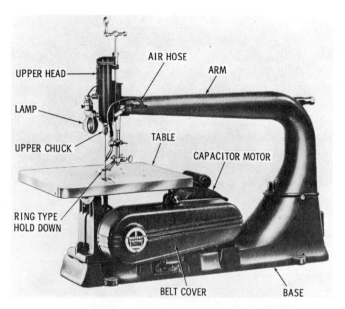

Fig. 2. The principle parts of a completely assembled jigsaw.

Fig. 3. The driving gears and table assembly of a typical jigsaw.

plished, shown in Fig. 4. The primary uses of the jigsaw are for cutting out intricate curves, corners, etc., as in wall shelves, brackets, and novelties of wood, metal, and plastics.

187

JIGSAW OPERATION

As previously mentioned, the jigsaw is used primarily for making various types of intricate wood cuttings which cannot readily be made on the band saw. The chucks are generally made to accommodate various sizes of blades, which are inserted with the teeth pointed in a downward direction. In operation, the front edge of the guide block should be in line with the gullets of the teeth and should be fastened in that position. The guide post is then brought down until the hold-down foot rests lightly on

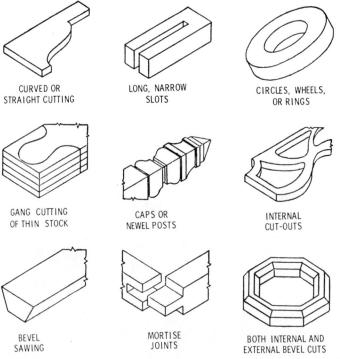

CURVED OR
STRAIGHT CUTTING

LONG, NARROW
SLOTS

CIRCLES, WHEELS,
OR RINGS

GANG CUTTING
OF THIN STOCK

CAPS OR
NEWEL POSTS

INTERNAL
CUT-OUTS

BEVEL
SAWING

MORTISE
JOINTS

BOTH INTERNAL AND
EXTERNAL BEVEL CUTS

Fig. 4. Various items produced by means of jigsaw cutting. One of the most important steps in cutting any shape from wood is that of marking pattern shapes on the wood to be cut. This is usually done by drawing the pattern with the aid of suitable squares or by the use of an outline projector. The pattern so produced can sometimes be mounted directly on the wood as a cutting guide, or else the pattern may be produced directly on the work by means of carbon paper.

the work to be sawed. For most work, the operation of the jigsaw does not differ in any important respect from that of the band saw.

The speed of the jigsaw is generally determined by the material to be cut, as well as the type of blade used, in addition to the skill of the operator. Various speeds of from 650 to 1700 cutting strokes per minute may be selected by the use of the proper step on the cone pulley. Jigsaw blades vary a great deal in length, thickness, width, and fineness of the teeth. All blades, however, may be grouped under two general classifications, which are as follows:

1. Blades which are gripped in both the upper and lower chucks.
2. Blades which are held in the lower chuck only.

The latter types of blades are known as saber blades, whereas the former are called jeweler's blade. The jeweler's blades are useful for all fine work where short curves predominate, while saber blades are faster cutting tools for heavier materials and medium curves. When it is desirable to make inside cuts, a starting hole is normally drilled at a suitable location.

SUMMARY

The jigsaw and band saw generally do the same type of work, although the jigsaw is more adaptable to sharp curves because smaller and finer blades are used. The jigsaw can cut inside holes since the blade is easily removed and inserted through a hole bored in the wood.

The speed of a jigsaw is generally varied by selecting the proper pulley groove, and is governed by the material used and the type of work to be done. The blades vary a great deal in length, thickness, width, and number of teeth per inch.

Blades are generally known as saber blades (which connect at only one end), and jeweler blades (which connect at both ends). Since the blade action is an up-and-down motion, sharper corners can be made and more accurate work can be accomplished.

REVIEW QUESTIONS

1. What is the motion action of the jigsaw blade?
2. What are some of the advantages in the jigsaw over the band saw?
3. Name the two types of blades used in the jigsaw.
4. What is the throat opening on a jigsaw?
5. How is the speed change on a jigsaw accomplished?

CHAPTER 13

Woodturning Lathes

A woodturning lathe is a machine tool used for shaping wood by causing the wood to revolve between centers while being acted upon by a sharp-edged cutting tool held in the hand and supported by a slide rest. Since the wood is revolving while being cut, the operation is termed *woodturning*.

The principal parts of a woodturning lathe, as shown in Fig. 1, are:

1. The bed.
2. The headstock.
3. The tailstock.
4. The tool post.

The bed usually consists of a heavy casting which resembles two parallel V-ways, and is supported by cast-iron legs on a table to bring the work up to the desired height. The two V-ways carry a rigid headstock at one end and a tailstock at the other, the latter arranged to slide on the V-ways and to be secured at any point by tightening the tailstock clamp. The headstock is bolted to the bed and contains the driving mechanism of the lathe. This consists of a hollow spindle supported between two bearings.

The spindle is revolved either by a step-cone pulley or by an individual motor in which case the spindle is directly connected to the motor. As shown in Fig. 2, a spur center is fitted into the spindle and engages one end of the wood to be turned, the other

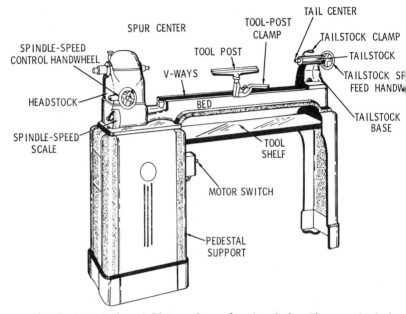

Fig. 1. A typical variable-speed woodturning lathe. The speed of the spindle on this particular lathe can be changed while the machine is running by means of a handwheel mounted on the headstock assembly. Spindle speeds from 300 to 2,600 rpm can be obtained when a 1740-rpm motor is used.

end being secured by the tail center. The spur center turned by the belt drive on one of the steps of the cone pulley and the spurs on the center cause the piece of wood inserted between the centers to turn also.

Both the tool post and the tailstock can be clamped to the bed at any point desired. By resting a sharp-edged cutting tool against the T-shaped tool rest, the wood is shaved off and the surface reduced to a circular form.

LATHE SPEEDS

The lathe speed may be regulated in several ways, depending upon the method of drive. When the lathe is driven from a set of step pulleys, the speed at the cutting edge will range from 1,000 to 2,500 feet per minute, depending upon the type of turning to be done. When the headstock is directly motor driven, the speed

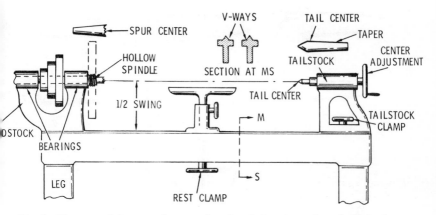

Fig. 2. The essential parts of a woodturning lathe. A section of the bed at MS shows the V-ways and V-grooved cut into the bottom of the headstock and tailstock.

is usually regulated by a rheostat or by a special switching arrangement that provides for up to four speeds, usually from 500 to 3,600 rpm. No definite rule for lathe speeds can be laid down, however, because of the large variations in diameter that often occur in the piece to be turned. Too slow a speed not only is a great waste of time, it will also leave a rough finish on the wood. Too fast a speed may damage the cutting tools. Always remember, the largest diameter of the material will determine the lathe spindle speed.

STARTING AND STOPPING THE LATHE

Prior to starting the lathe, the adjustments and clamps should be tested to assure their workability, and the work should be revolved by pulling the belt by hand to make sure that the work clears the tool post. It is better to start turning at a safe speed and increase it, if necessary, after the work is rough-turned and is running true.

The lathe should never be run at high speed if any part of the work is off center or out of balance. If the work to be turned is out of balance, it should be counter-balanced by suitable weights, although this may not always be possible. To stop the machine, the belt is shifted, or the motor is disconnected and braking is accomplished by placing the hand on the pulley. Large faceplate

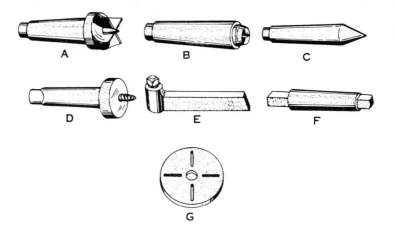

Fig. 3. Lathe accessories: (A) Spur drive center; (B) Cup center; (C) Cone center; (D) Screw center; (E) Tool holder; (F) Adapter; (G) Faceplate.

work, however, should not be stopped too suddenly in this way, as there is danger of unscrewing the plate from the spindle.

A smooth piece of work can be safely stopped by braking on the work itself, using a handful of shavings or smooth turnings to prevent burning. However, the lathe should never be stopped by placing the hand on the back or edge of the faceplate, since there are likely to be projecting objects on it that may cause injury to the operator.

LATHE ATTACHMENTS

Lathe attachments (Fig. 3) consists of several tools or accessories necessary to properly perform the work. They are: *lathe centers, drivers, screw chucks, center chucks, faceplates,* and various *turning* and *measurement* tools.

Lathe Centers

The function of the lathe centers is to hold the work and revolve it between the spindles. They are of three principal types—the *spur* or *live center,* the *headstock* or *tailstock center,* and the *cup center.*

The spur or drive center is used in the live spindle for driving small and medium sized pieces that are to be turned between the centers. The spur center is tapered to fit the hole in the spindle, and the driving end has a point and either two or four spurs to engage the wood. The spur center is inserted in the piece to be turned by setting the point in the center and driving the spurs in by striking the end of the center with a mallet. The center is removed from the lathe by pushing a rod through the hole in the spindle.

The tailstock center is used to support the right-hand end of the work and is generally self-discharging. It may be removed from the spindle by drawing the spindle back with the handwheel as far as it will go. This action will automatically push the center out of the spindle.

The cup center is also a tailstock center and has a thin circular steel edge around a central point. It is more accurate than the cone center and does not split the wood so easily. Since the dead or tailstock center does not revolve, the end of the stock which turns on them should be well oiled to prevent burning by friction.

Lathe Drivers

Drivers are devices, other than the spur center, used for revolving the work. There are several forms available, with some made to be used with a small slotted faceplate and dog (as in machine-shop turning), the center plate having a projection to which the dog is fastened. Another form of driver is one in which the center has a square shank over which the dog fits.

Screw centers are small faceplates with a single screw in the center, and are used for turning small pieces. Some screw centers hold the central screw firmly and has an attachment for regulating the projection of the screw through the face.

Faceplates are holding devices used for work which cannot readily be driven by any of the foregoing methods. They contain screw holes countersunk on the back for the screws used in fastening the wooden face or chuck plates. The smaller sizes are frequently provided with a recess in the center for rechucking purposes. They are made in various sizes, depending upon the size of the lathe and the stock they must support. Objects such as circular disks, bowls, trays, etc., are always turned on faceplates.

Woodturning Tools

Woodturning tools (Fig. 4) are used for cutting and scraping purposes, the most common of which are various types of gouges,

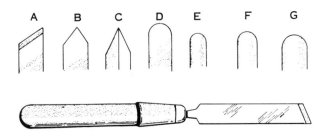

Fig. 4. Woodturning tools: (A) Skew chisel; (B) Spear-point chisel; (C) Parting tool; (D) Round-nose chisel; (E) Gouge; (F) Gouge; (G) Gouge.

chisels, and parting tools. Gouges are beveled on the outside or convex side, and the length of the bevel is about twice the thickness of the steel. Skew chisels are beveled on both sides, and their cutting edges form a 60° angle with one side of the chisel. Parting or cut-off tools are thicker in the center of the blade than at the edges and, thus, will not bind or overheat when a cut is made. This tool is used for making narrow cuts to a given depth or diameter.

Measurement Tools

Other necessary tools are those used for measurements (Fig. 5), usually termed sizing and layout tools. The tools for sizing and layout work on the lathe include both inside and outside calipers, dividers, and trammels. The latter should be of a type fitted with both outside and inside caliper points as well as the ordinary trammel points. Measuring rods are also used for sizing faceplate work, and are used for sizing both the inside and outside of a ring-formed object.

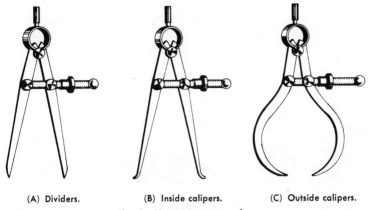

(A) Dividers.   (B) Inside calipers.   (C) Outside calipers.

Fig. 5. Measurement tools.

MEASUREMENTS

The ability to take accurate measurements plays an important part in woodturning work, and can be acquired only by practice and experience. All measurement should be made with an accurately graduated scale.

An experienced operator can take measurements with a steel scale and calipers to a surprising degree of accuracy. This is accomplished by developing a sensitive caliper feel and by carefully setting the calipers so that they split the line graduated on the scale.

Outside Calipers

A good method for setting an outside caliper to a steel scale is shown in Fig. 6. The scale is held in the left hand and the caliper in the right hand. One leg of the caliper is held against the end of the scale and is supported by the finger of the left hand while the adjustment is made with the thumb and first finger of the right hand.

The proper application of the outside caliper when measuring the diameter of a cylinder or a shaft is shown in Fig. 7. The caliper is held exactly at right angles to the center line of the work, and is pushed gently back and forth across the diameter of the cylinder to be measured. When the caliper is adjusted properly, it should

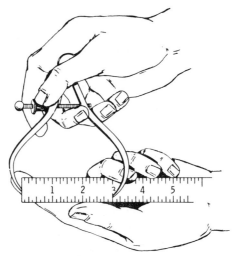

*Fig. 6. Method of setting an outside caliper to a
steel square.*

slip easily over the shaft of its own weight. Never force a caliper
or it will spring and the measurement will not be accurate. Never
grip the caliper too tightly; the sense of touch will be very much
impaired.

Inside Calipers

To set an inside caliper for a definite dimension, place the end
of the scale against a flat surface and the end of the caliper at the
edge and end of the scale. Hold the scale square with the flat surface. Adjust the other end of the caliper to the required dimension,
as shown in Fig. 8.

To measure an inside diameter, place the caliper in the hole,
in the position shown in Fig. 9 by the dotted line, and raise the
hand slowly. Adjust the caliper until it will slip into the hole with
a very slight drag. Be sure to hold the caliper square across the
diameter of the hole.

In transferring a measurement from an outside caliper to an
inside caliper, the point of one leg of the inside caliper rests on a
similar point of the outside caliper, as shown in Fig. 10. Using this

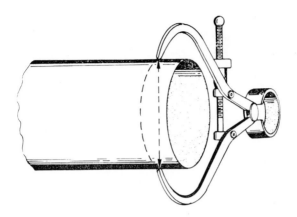

Fig. 7. Method of measuring with an outside caliper.

contact point as a pivot, move the inside caliper along the dotted line shown in the illustration, and adjust with the thumb screw until you feel the measurement is just right. The hermaphrodite caliper shown in Fig. 11 is set from the end of the scale exactly the same as the outside caliper.

The accuracy of all contact measurements is dependent upon the sense of touch or feel. The caliper should be delicately and lightly held in the finger tips—not gripped tightly. If the caliper is gripped tightly, the sense of touch is lost to a great extent.

CENTERING AND MOUNTING STOCK

Wood stock to be turned must be properly marked; that is, the true centers must be obtained prior to turning. In square stock, the center is usually determined by the diagonal method, which consists of holding the center head of a combination square firmly against the work and drawing two lines at right angles to each other and close to the blade across each end of the work.

After the work center has been marked, the work can be mounted in the lathe as shown in Fig. 12. The stock is first pressed against the spur or live center so that the spurs enter the grooves previously marked. Next move the tailstock up to about 1 inch from the end of the stock and lock it in this position. Then advance

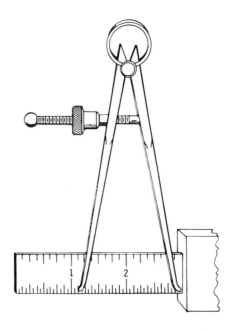

Fig. 8. Method of setting an inside caliper.

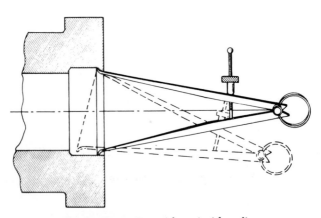

Fig. 9. Measuring with an inside caliper.

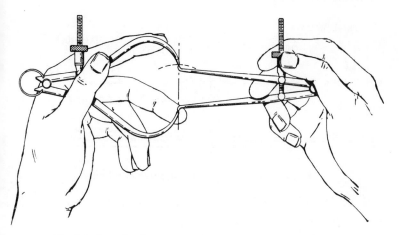

Fig. 10. Transferring a measurement from an inside to an outside caliper.

the tailstock center by turning the feed handle until the center makes contact with the work. Continue to advance the center while slowly rotating the stock by hand. After it becomes difficult to turn the stock, slack off on the feed about one-quarter turn and lock the quill spindle. The stock is now ready for turning.

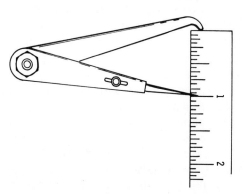

Fig. 11. Setting a hermaphrodite caliper.

The tool rest is next mounted in position so that it is level with the centers and about 1/8-inch away from the stock. Clamp the

201

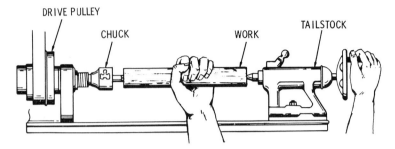

Fig. 12. Method of mounting work in a lathe.

tool rest in position and revolve the stock by hand to make certain it has sufficient clearance.

TYPICAL WOODTURNING OPERATIONS

The most common woodturning operations include such simple steps as roughing off, paring or finishing, squaring of ends, making concave or ring cuts, etc. In a roughing-off cut, place a large gouge on the rest so that the level is above the wood and the cutting edge tangent to the "circle of cut." In adjusting the rest for this, the handle of the gouge should be well down. Roll the gouge over slightly to the right so that it will shear instead of scrape the wood. Lift the handle slowly forcing the cutting edge into the wood. Remove the corners of the wood first, then the intervening portions. The tool should be held at a slight angle to the axis of the wood being turned, with the cutting end in advance of the handle, as shown in Fig. 13.

In making a paring or finishing cut, lay a skew chisel on the tool rest with the cutting edge above the cylinder and at an angle of about 60° to the surface. Slowly draw the chisel back and, at the same time, raise the handle until the chisel cuts about 1/4 to 3/8 inch from the heel. Begin the first cut about 1 to 2 inches from either end, pushing toward the other end. Then begin at the first starting point and cut toward the other end, thus taking off the rings left by the gouge, and cutting down to where the scraped sections are just visible. Take a last cut to remove all traces of the rings and to bring the wood down to a uniform diameter from end

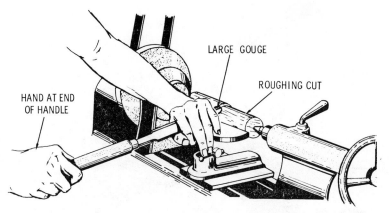

Fig. 13. The correct way to hold a gouge in making a rough cut on the lathe.

to end. To test for smoothness, place the palm of the hand, with the fingers extended, lightly on the back side of the wood away from the tool rest.

After the work has been reduced to a cylinder, the ends may be squared to mark the end of the turning. This is commonly accomplished by placing the 1/4- or 1/2-inch skew chisel on the tool rest and bringing the cutting edge next to the stock perpendicular to the axis, as in Fig. 14. The heel of the chisel is then slightly tipped from the cylinder to give clearance. Raise the handle and push the chisel toe into the stock about 1/8-inch outside the line indicating the end of the cylinder. Swing the handle still further from the cylinder and cut a half V to give clearance and prevent burning. Continue this process until the cylinder is cut through to about a 1/16-inch diameter.

When making concave cuts, place the gouge on the tool rest with the cutting edge well above the wood. Roll the tool to the side until the bevel at the cutting point is perpendicular to the axis of the cylinder. Slowly raise the handle to force the gouge into the wood. When the gouge bites into the wood, force it forward and upward by slightly lowering the handle while rolling it back toward its first position. Reverse the position of the gouge and make a similar cut from the other side to form the other half of the semicircle.

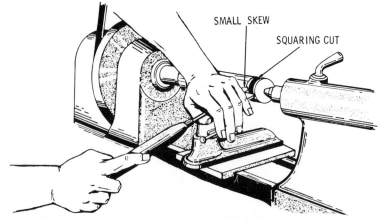

Fig. 14. *A skew chisel is held in this manner to square the ends of the work.*

Ring cuts are made by placing a 1/4- or 1/2-inch small skew chisel on the tool rest with the cutting edge above the cylinder and the bevel tangent to it. Draw the chisel back and raise the handle, bringing the heel of the chisel in contact wtih the cylinder at the marked line. The chisel is then moved to the right (if cutting the right side of the bead) while being continually tipped to keep the lower bevel tangent to the revolving cylinder and to the head at the point of contact. Continue the cut until the bottom of the convex surface is reached.

POLISHING

The polishing of turned work in the lathe is a great deal easier than applying polish to a flat surface. Here, the polishing operation is done with a cloth while the work is rotating in the lathe and after it has been sanded.

In sanding, first use a fairly coarse grade of paper (No. 1 or 1-1/2) followed by a fine grade (No. 0 or 00.) Before applying the polishing cloth, the wood may be varnished lightly while the lathe is not running, taking care to wipe off all surplus varnish. The varnish will assist in giving the surface a fine polish when the cloth is applied.

SUMMARY

The principal parts of a woodturning lathe are the bed, headstock, tailstock, and the tool post. The headstock and the tailstock support the wood which is being turned. The tool post supports the chisel or gouge which cuts or shapes the wood being turned.

There are many woodturning tools, both for cutting and for measuring. Cutting tools include skew chisels, spear-point chisels, parting tools, round-nose chisels, and gouges. The skew chisel is beveled on both sides, and forms a cutting edge of 60°.

REVIEW QUESTIONS

1. Name the four main parts of the cutting portion of a woodturning lathe.
2. What is the difference between inside calipers and outside calipers?
3. What is the difference between dividers and calipers?

CHAPTER 14

Planers, Jointers, & Shapers

PLANERS

The planer is a machine for planing or "surfacing" wood by means of a rapidly revolving cutter which chips off the rough surface in many shavings. The piece to be planed is passed over (or under) the revolving cutter by power feed, leaving a smooth or finished surface. The planer is thus used mainly for surfacing flat stock and for reducing stock to a uniform thickness.

These machines are available with either one or two cutting heads so that one or both sides of a piece of lumber may be planed in one operation. The 24- and 30-inch planers, suitable for fairly wide panels, are most commonly used in the wood shop, although faster machines for narrower material (Fig. 1) are typical of those in planing mills. Warped or twisted stock should always be straightened on one surface with a jointer before being put through the planer.

Construction

The essential parts of a planer include a heavy casting or *frame, table or bed, feed rolls, cutter head, chip breaker,* and *pressure bar.* The table can be raised or lowered to accommodate the thickness of the stock to be planed.

Two smooth steel feed rolls are mounted in the table or bed casting, and project slightly above its upper surface, as shown in

Courtesy Parks Woodworking Machine Co.
(A) A 20-inch single-surface
planer.

Courtesy Northfield Foundry & Machine Co.
(B) A No. 7 Northfield single-surface
planer.

Fig. 1. Typical thickness planers.

Fig. 2. Two additional feed rolls are mounted directly above these on top of the frame. The one rear feed roll is smooth like those in the bed, whereas the feed roll in the front is corrugated in the lengthwise direction. The function of the feed rolls and accompanying mechanism is to control the feed of the work to be planed. The feed rolls are mounted in pairs and are usually interconnected by gears. The corrugated feed roll is sometimes mounted in one piece, although on modern planers they are commonly mounted in sections, each section being equipped with inside springs which permit the rollers to adjust themselves to the varying thickness of the work.

On small planers, the feed mechanism is driven by the same motor as the cutter head, but on large machines the feed mechanism is driven by an independent motor. The rate of feed may generally be adjusted to a value of from 20 to 130 feet per minute. Slow feeds are used for very smooth finish work.

The cutter head (Fig. 3) consists of two or more cutting knives set into slots cut lengthwise along a steel cylinder. This cylinder (cutter head) revolves at a speed of from 3,600 to 7,200 rpm, and is usually mounted in large precision ball bearings housed squarely over the frame.

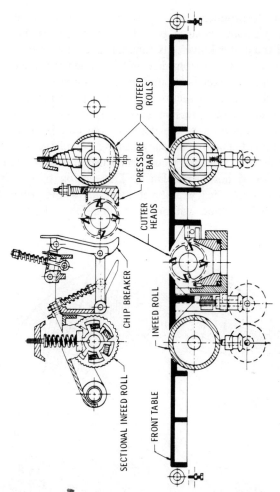

Fig. 2. Principal parts of a typical thickness planer.

The wood to be planed is supported on the tables and is pushed across the revolving cutter head. Modern cutter heads are generally equipped with four cutting knives set at a 20° or 30° cutting angle. Screws for adjusting the knives to the cutting circle are so designed that they will not bind, and they fit the holes very closely, thus preventing dust from working down and clogging their movement.

209

In order to produce a smooth surface, all planers are equipped with a device known as a *chip breaker,* which is mounted just ahead of the cutter knives. Its function is to hold the wood firmly to the bed and keep it from driving back when the knives start the cut. The chip breaker is usually made up of several sections and designed so that each section works independently with a yield against a heavy compression spring. This spring tension is increased in proportion to the depth of the cut. Behind the cutter head is mounted another heavy casting called the pressure bar. Its function is to hold the wood firmly to the bed or center platen after the wood has been planed. The tension of this bar must be comparatively heavy and yet must allow the wood to carry through the machine without stopping. Each type of planing

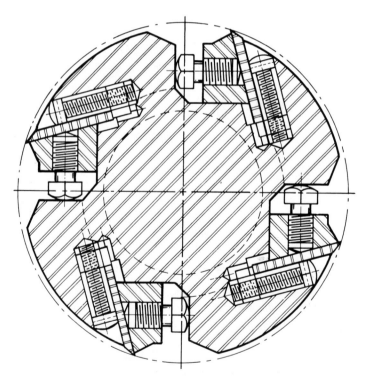

Fig. 3. Cross-section of a cutter head used on a thickness planer.

210

requires a different adjustment of this bar, so the method of adjusting must be simple and accurate to the thousandths of an inch. The first setting is usually made with the fine-thread adjusting screws at each end of the bar. These screws are fitted with dial indicators marked for direction *up* and *down* and are locked into position with a spring lock.

Some planers are equipped with a grinding and truing device which slides on a bar above the cutter head. The grinding is performed with a small, high-speed motor, while the truing is done with a stone as the cutter head revolves at full speed.

Adjustments

The planer is a highly complex machine requiring many different adjustments, among the most important of these being the cutting angle of the knives. The cutting angle may be altered by changing cutter heads or by grinding a back bevel on the knife edge.

The pressure bar should be set parallel with the cutting circle of the knives, using a good parallel block about 4" square. The pressure-bar springs are adjusted for varying tension on the stock. Arrows show the direction to turn the screws for moving the bar up or down. For a first cutting, or after grinding the knives, it may be necessary to raise this bar slightly, perhaps one or two notches. Care should be used to keep the bar parallel.

Most modern machines are equipped with a quick-acting micrometer central control. This control should be set .005" above *Low* when checking the pressure bar or setting the bar in line with the cutting circle of the knives. Moving the control level toward *High* raises the bar from 0 to .040". This range will usually take care of all pressure-bar adjustments from first cutting to finishing work, and also during many knife grindings. The cam mechanism and bearings should be kept well oiled. Adjustment of the chip breaker is made by means of an adjusting screw under the center arm. Adjust the chip breaker even with (but not lower than) the cutting circle of the cutter head, thereby giving each finger of the bar full variation.

The sectional in-feed roll should occasionally be checked; keep it 1/16" (3/32" for rough stock) below the cutting circle. Keep only the necessary tension on the in-feed roll box springs to carry

211

the stock through. This allows free action of the entire roll. Occasionally, introduce a little thin oil over the top of the entire sectional roll. Do not tighten the box springs so the entire roll cannot act.

An efficient sectional roll is a great aid to production planing, especially on narrow stock. When the outer sections do not maintain their normal central position, it is evident the springs need replacing. Remove the roll with the driving gear from the machine and remove the collar from the end opposite the gear. Strip all the sections from the shaft, clean, respring, and pack with grease. Replace each unit on the shaft so that the shaft drives the inner section against the lugs and not against the springs. Place a dividing disc or separator between each unit and tighten the collar against the shoulder of the shaft.

The out-feed roll should be checked occasionally; keep it 1/16″ (never more) below the cutting circle. Keep strong tension on the top out-feed roll box springs. Keep the machine clear of knots and chips if a weak exhaust system is used.

For finish work, the lower rolls are set so that a sheet of paper (.003″ to .004″) can be drawn from between a straightedge and the center table, with the straightedge resting on both lower rolls. When adjusting the lower rolls, first loosen the steel plates over the boxes to avoid cramping them. Machines equipped with a simultaneous micrometer adjuster for the lower rolls have automatic spring tension over the roll boxes. For material that is rough on both sides, set the rolls from .006″ to .010″ above the center table.

When simultaneous micrometer adjustments for the lower feed rolls are necessary, first loosen the small locking knob under the front table. Set the finger wheel at .005″ on the dial and tighten the small locking knob. Check the lower rolls, following the instructions for setting, except set them at .005″. All future setting of the lower rolls is then done by means of the finger wheel. The reading on the dial indicates the height in thousandths of an inch the rolls are above the center table; .004″ for smooth panel-finishing work; .005″ to .007″ for narrow finishing work; .008″ to .010″ or more for rough work. Soft wood requires high roll setting; very hard wood, a very low setting.

The cutter-head knives are set 3/32″ from the lips and should be reset when worn to 3/64″. To reset the knives, turn the jack screws in (clockwise) about one and one-half turns, or just enough to allow the adjusting or lifting screws to raise the knife. Too much looseness will prevent accurate adjustment. Adjust the knife to the hand-setting attachment that is made to raise the smooth part of the cutter head, or to the disc-setting device used in the jointer head. The former is more convenient. Tighten the knife just enough to hold it firmly in place and proceed to the next knife.

When all knives are properly set, tighten them all evenly and securely, proceeding around the cutter head rather than lengthwise. Do not attempt to adjust one part of the knife with another part held fast. This puts a strain on the knives and will either drive them back or cause them to break in service. The adjusting screws should not have undue strain put upon them. If the knives are driven back, it would be desirable to clean the chip-breaker bars and oil the threads of the jack screws.

Operation

After adjusting the machine to the thickness of the work to be planed (the planer table should be set to approximately 1/16″ less than the thickness of the work), the cutter head and feed rolls are started in the conventional manner. The work is fed in between the rolls of the machine in such a way that the knives will cut with the grain and not against it. If a great amount of material must be removed from the work, it is customary to plane off an equal amount on both sides; this is necessary in order to prevent the work from warping because of an unequal amount of moisture content.

JOINTERS

The jointer is similar to the planer, except that it has only one cutter head located below the table, and the feed is usually by hand. Its primary use is to cut a true face and edge on stock that is warped, twisted, or has other irregularities. The table is in two sections, the in-feed section being raised or lowered to control the thickness of the cut. The out-feed part of the table is raised

213

or lowered by a unit to control the thickness of the finished piece. By tilting the fence, bevel edges can be cut. This machine can also be used for tapering, end planing, and rabbeting.

Construction

The essential parts of a jointer (Fig. 4) consists of a *base* or frame supporting the assembly, *cutter head, tables or bed, guide fence,* and *table-adjusting hand wheels.*

The cutter head houses two or more high-speed steel knives, and revolves between the front or in-feed table and the rear or out-feed table. The tables may be raised or lowered by means of hand wheels to regulate the depth of cut.

A guide fence extends along both tables to hold the work as it is advanced toward the cutting knives. The fence may be tilted

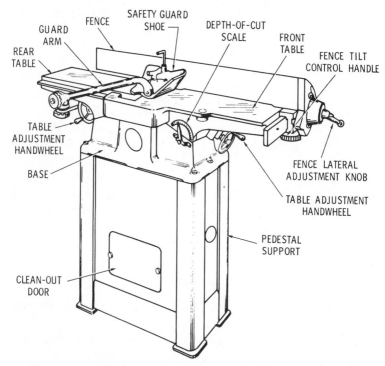

Fig. 4. A 6-inch jointer equipped with a dual-purpose safety guard and hold-down shoe.

45° either way and may also be moved crosswise along the jointer table. The cutter head is generally provided with a guard cover as a safety measure.

The rotating speed of the cutter head is usually between 3,600 and 6,000 rpm, although for certain types of work a speed of up to 8,000 rpm may be used. Since the speed of the conventional 60-cycle ac motor is 3,600 rpm or less, various types of pulley ratios are employed to obtain this higher speed.

Adjustments

One of the most important adjustments on the jointer is the relation of the rear table to the cutter head. To do satisfactory work, this table must be exactly level with the knives at their highest point of rotation. To make this adjustment, instructions issued by the manufacturer of the equipment should be followed. Fig. 5 shows the correct and incorrect positions of the rear table.

The in-feed table, on the other hand, must be somewhat lower than the rotating knives, the exact height of which depends upon the depth of the cut. The difference in height between the two tables is equal to the depth of the cut. Improperly adjusted tables will result in the work being cut on a taper.

When knives become dull or nicked, it is necessary to either replace them or sharpen them. Jointing knives may be sharpened and brought to a true cutting circle by jointing their edges while the cutter head is revolving. In performing this operation, a fine-cutting oilstone is placed on the out-feed table, and the table lowered until the knife just touch the stone at their highest point of rotation.

There are numerous other methods used to sharpen jointer knives, each having its particular advantage depending on the knowledge and experience of the operator or sharpener. When it becomes necessary to replace or reset the knives in the cutter head, the manufacturers' instructions should be followed, since different cutter heads have different knife-holding methods.

Operation

Although a great number of woodworking operations may be performed on the jointer, the most common are edge planing,

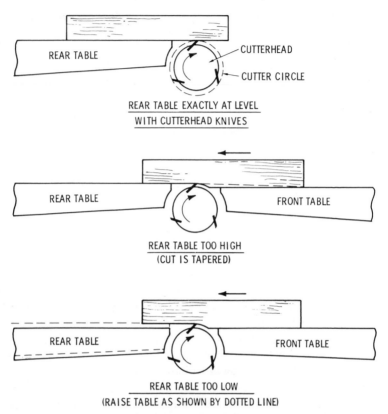

REAR TABLE EXACTLY AT LEVEL
WITH CUTTERHEAD KNIVES

REAR TABLE TOO HIGH
(CUT IS TAPERED)

REAR TABLE TOO LOW
(RAISE TABLE AS SHOWN BY DOTTED LINE)

Fig. 5. Showing the effect of rear-table adjustment on the jointer. The rear table must be adjusted until it is exactly level with the cutter knives at their highest point of rotation. The front table is lowered by the amount of the desired cut.

planing of ends, and surfacing. Among these, edge planing or jointing an edge is the simplest and most common operation. Assuming the machine to be properly adjusted, and the guide fence square with the table, all that is necessary is to feed the work over the rotating cutter heads. The left hand presses the work down on the rear table so that the cut surface will make a good contact with the table. The right hand exerts no downward pressure on the front table, but simply advances the work over

216

the knives. Both hands, however, exert pressure on the work to force it against the guide fence to obtain the proper edge.

Edges on thin work, such as veneer, may also be planed on the jointer provided the work is held firmly between two suitable boards or planks. In most edge planing operations of this type, special clamps are used to hold the work in place.

Planing across the ends of boards may be performed in a manner similar to planing with the grain. The necessary precaution consists in taking only light cuts each time because of the tendency of the grain at the end of the board to split if thick cuts are attempted. To avoid splitting the end wood, some operators make only a short cut at one edge, then reverse the wood and complete the planing from the opposite edge.

Surfacing the faces of a board is one of the more difficult jointer operations. When planing the faces, the grain pattern in the

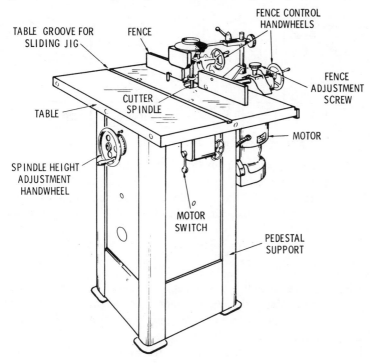

Fig. 6. A typical shaper.

edges should be noted. To plane the board with the grain, the two sides of the board must be planed in opposite directions.

SHAPERS

A shaper (Fig. 6) is used mainly to place an edge on stock having an irregular outline, and for rabbeting, grooving, and fluting. Some of the cuts that can be made are shown in Fig. 7.

A shaper consists essentially of one or more vertical spindles with cutting heads, table, fence, and base. Generally, a knife with its cutting edge ground to the desired shape of cut is used, but solid cutters milled to the desired shape are also available.

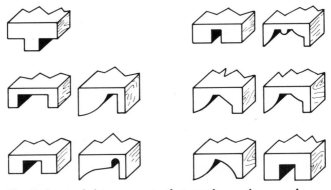

Fig. 7. Some of the many cuts that can be made on a shaper.

Stock may be cut roughly to shape with a band saw or jig saw before being finished on the shaper. When a number of pieces of the same pattern are to be shaped, special templates or forms are often used to regulate the depth of cut by bearing against the shaper collar.

Construction

Shapers are usually constructed with one or more vertical spindles which carry the cutter heads. Two spindles revolving in opposite directions are perhaps the most general and useful form, since one cutter can cut with the grain of wood running in one direction and the other cutter can operate on wood with the grain running in the other direction without stopping to reverse the machine.

The spindle (Fig. 8) projects above the surface of an accurately planed table. Cutters are attached to this spindle, and the assembly rotates at speeds of 5,000 to 10,000 rpm or higher. On account of this high speed, pulleys of about three-to-one ratio are generally necessary when the machine is driven by a standard 3,600 rpm motor.

A wide variety of knives, saws, collars, etc., are used in shaper operations. Shaper knives are of two kinds—the open-face or flat knives, and the three-lip cutters. The flat knives are used mainly on large machines, and have beveled edges which fit into

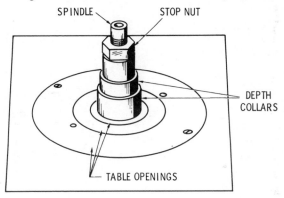

Fig. 8. The shaper spindle projects through the table.

corresponding slots milled in two collars. The knives of exactly the same width are always used, and they must be clamped securely to prevent them from loosening while the machine is in operation.

The three-lip shaper cutters have three cutting edges, and slip over the spindle shaft. This type of cutter is much safer to work with. Fig. 9 illustrates some of the various types of cutters available.

Operation

There are four main methods used to hold or guide the stock against the shaper knives:

1. With guides or fences.
2. Against collars.
3. With an outline pattern.
4. With forms.

Each one of these methods is widely employed and each is suitable for a certain kind of work. Fig. 10 illustrates the use of these four methods. In Fig. 10A, the work is held against a fence as it is advanced into the cutter. The position of the fence thus determines the depth of cut. This method is used only for straight cuts. (Adjustment of a shaper fence is shown in Fig. 11.) In Fig.

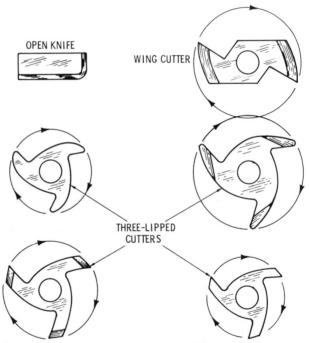

OPEN KNIFE

WING CUTTER

THREE-LIPPED
CUTTERS

Fig. 9. Various types of cutters are available for use on a shaper.

10B, the depth of cut is regulated by a collar—a deeper cut requires a collar with a smaller diameter. The edges of an irregular-shaped piece of material can be cut using this method. The use of an outline pattern is shown in Fig. 10C. With this method,

220

the pattern rides against the collar to determine the depth of cut. Fig. 9D shows a form being used to hold the work in position so that it can be advanced to the shaper cutter.

Where a job calls for working the stock directly against an ordinary shaper collar (such as in Fig. 9), the friction created between the collar and the work has a tendency to burn and mar the material. This burning can be avoided, however, by using a ballbearing guide collar in place of the regular type. The outer shell of the ball-bearing collar revolves independently of the

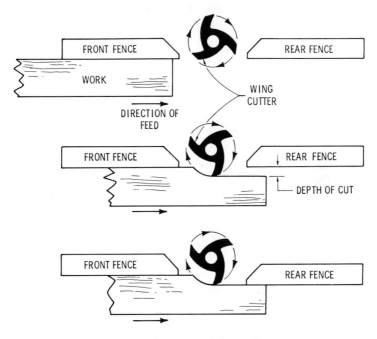

Fig. 10. Adjustment of shaper fences.

inner shell, and thus rolls along with the feed, practically eliminating all wear and friction on any forms or guide patterns being used or on the material being shaped.

Heavy cutting is best accomplished with open-faced knives set in open collars, with the asembly being well balanced to avoid vibration to render the set-up as safe as possible. Whenever it is necessary to work directly against the shaper collar, it is well to rest the stock against a steel guide-pin, while easing it slowly to the knives at the beginning of each cut. If the character of the work is such that it is impossible to employ a hold-down device, two or more light cuts should be made where considerable material is to be removed. It is advisable to relieve the cut as much as possible by chamfering off the corners of the stock with a band saw on which the table can be tilted, or on a tilting-blade bench saw. The equipment which is really best adapted to extra-heavy

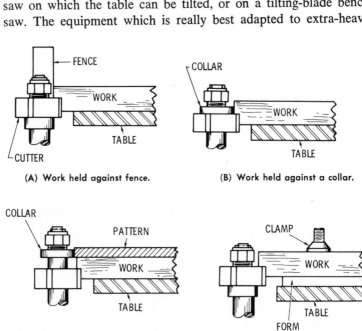

(A) Work held against fence.   (B) Work held against a collar.

(C) Use of a pattern and collar.   (D) Work held by a form.

Fig. 11. Four principal methods used to hold the work against the shaper knives.

cutting is the type of shaper with over-head bearings, although some operators prefer a machine having an auxiliary drop table.

In the type of operations employed in the average mill-work shops, shaping with guides is a very popular method. The most common types of guides are various types of fences to suit the particular operation desired. Capable operators keep an ample supply of fences on hand, each usually being adapted to one particular class of work. These generally include one for chamfering, another for grooving, and still others for panel raising, rabbeting, and other set-ups called for most frequently. When using a saw on the shaper spindle for grooving, it is customary to tack a strip of plywood panel stock on the face of the fence, which is then positioned so the blade cuts through the plywood the proper distance to make the required depth of groove.

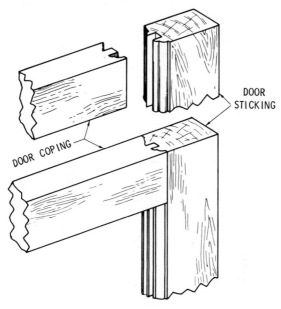

Fig. 12. A shaper is used to form the parts of a door.

When milling short and light work, the strips of hardwood which serves as a guard and hold-down can sometimes be employed as a fence. This is because it is readily set at any required distance from the cutters by means of the adjusting screws.

When machining curved and irregular-shaped parts, some operators use small curved fences equipped with adjusting rods which fit into the regular guard and hold-down device. They usually find this method better than working the stock against the shaper collar, as the friction produced by the latter method frequently burns and glazes the stock.

223

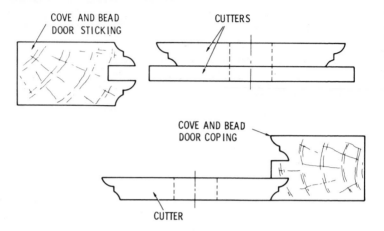

Fig. 13. The same cutter is used to form both mating sections of a door.

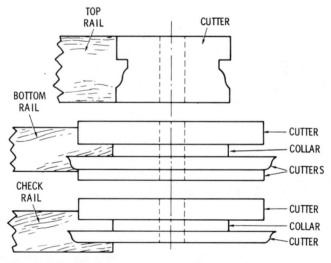

Fig. 14. Shaper cutters used to form the parts of a window sash.

Power and Speed

The size of the motor necessary to operate a shaper depends upon the type of work to be done. The medium sized shaper using 1/2″ hole cutters work satisfactorily with a 1/2-horsepower motor. In cases where large knives mounted between slotted collars are used, a 3/4- to 1-horsepower motor will be necessary.

The motor must be a 3,450-rpm type in order to give the shaper spindle the required speed. With a pulley ratio of 3-to-1, the actual spindle speed will be in the order of 10,000 rpm. The motor should preferably be of the reversible type, since an opposite direction of rotation may often be required.

A few of the many shapes that are cut by using a shaper are shown in Figs. 12, 13 & 14.

SUMMARY

A planer is a machine used to plane wood by means of removing the rough surface. The wood to be planed is passed under the cutting heads, leaving a smooth or finished surface. Many machines are available with two cutting heads so that both sides of a piece of lumber may be planed in one operation. A planer is generally power fed which means the lumber is automatically fed into the cutting heads.

The jointer is very similar to the planer, except it has only one head generally located below the table. The lumber is usually fed by hand. The primary use of the jointer is to cut a true surface on lumber that is warped or twisted.

A shaper is used to finish the edge on stock lumber, and for rabbeting, grooving, and fluting. Many times, stock may be cut roughly to shape with a band saw or jig saw before being finished on the shaper.

REVIEW QUESTIONS

1. Explain the difference between a planer and jointer.
2. What are the principal operations of a shaper?
3. How many cutting heads are used on a planer?

CHAPTER 15

Mortisers and Tenoners

The laborious operation of cutting a mortise-and-tenon joint by hand has been overcome by the use of various types of mortising machines now available. The mortise-and-tenon joint has always been recognized as one of the best methods of joining the end of one piece of wood to the side of another, and is therefore widely used. Tenons can be made on a special tenoning machine, or circular saws and band saws, whereas mortises can only be made on mortising machines or machines equipped with mortising attachments.

MORTISERS

There are several types of mortising machines on the market, although the hollow-chisel machine is most commonly employed in the average mill or woodworking plant. Fig. 1 shows an elementary type machine in which an auger bit having no lead screw revolves at high speed inside a hollow chisel. This hollow chisel has four sides of equal width and, as it is forced into the wood, the bit bores a hole which is cut square by the sharp end of the chisel. Thus, the chisel and the bit operating together accomplish what is equivalent to boring a square hole. That is, the hole is bored and squared at the same time. By shifting the work and repeating the operation, a mortise is produced with square ends and a square bottom, and of the desired length. The hollow chisel and auger bit is shown in Fig. 2.

In operation, the tool is forced into the work by pressure· on a foot lever, a return spring bringing it back to the initial position

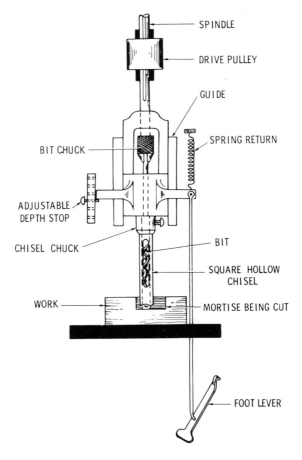

SPINDLE

DRIVE PULLEY

GUIDE

SPRING RETURN

BIT CHUCK

ADJUSTABLE
DEPTH STOP

CHISEL CHUCK

BIT

SQUARE HOLLOW
CHISEL

WORK

MORTISE BEING CUT

FOOT LEVER

**Fig. 1. An elementary machine showing the principles
of a hollow-chisel mortiser.**

when the pressure is released. The depth of the mortise is regulated by an adjustable depth stop.

The chisel capacity of this type of mortiser is from 1/4" to 3/4" square, although a power-feed machine is recommended for the larger size where parts are mortised in quantity. Chisels with 2-3/4" and 4" blades are available. with an adjustable stroke up to four inches. Provisions are made on the power-feed machines for quickly adjusting the stroke and leverage to suit the size

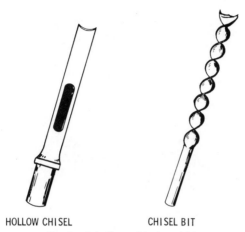

HOLLOW CHISEL   CHISEL BIT

Fig. 2. A hollow chisel and bit.

of chisel and depth of mortise. This permits the operator to use the best feed for the work.

TENONERS

There are several types of tenoners and they may be classified with respect to the cutting tools and their motion as:

1. Reciprocating planes.
2. Circular saws.
3. Rotary cutters.
4. Hollow auger.

The first mechanical device for cutting tenons employed an arrangement of circular saws working at right angles to each other. One pair of saws running parallel to the grain of the wood cut the sides of the tenon, and another pair running at right angles cut the shoulders. The two saws cutting the sides of the tenon were of the same diameter and were mounted on the same spindle, a collar being placed between them to regulate the thickness of the tenon. The saws cutting the shoulders were mounted on separate spindles running at right angles to the pair of saws which cut the sides.

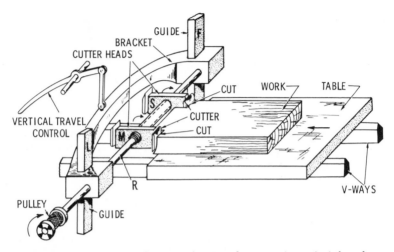

Fig. 3. An elementary diagram showing the operating principles of a rotary cutter tenoner.

An elementary tenoner illustrating the work principles of a rotary-cutter type is shown in Fig. 3. Here, two cutter heads, M and S, similar to planer cutters are placed on a spindle R, whose journals are carried by a sliding bracket having a vertical movement along the guides L and F. The cutter heads may be placed any distance apart, corresponding to the thickness of the tenon to be cut, by placing a collar between them whose length equals the thickness of the tenon. The table slides on V-ways at right angles to the spindle so that when the work is clamped on the table it may be brought into position under the cutter for the desired length of tenon.

With the cutters raised, the work is pushed under the cutters in the direction of the arrow to the desired length of the tenon to be cut. The cutters, which are revolving rapidly, are slowly passed across the work by lowering the bracket with the vertical control, thus planing off the sides and forming the tenon, no separate operation being necessary for cutting the shoulder.

Various types of mortise and tenon joints are illustrated in Fig. 4, while Fig. 5 shows some of the cuts and joints made on more elaborate tenoning machines.

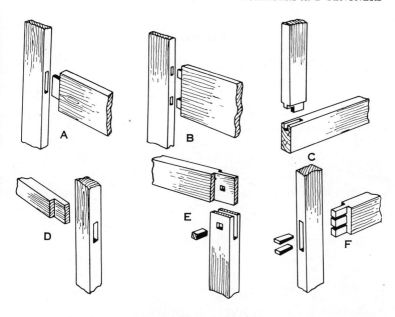

Fig. 4. Various mortise and tenon joints; (A) Single mortise and tenon;
(B) Double mortise and tenon; (C) A haunched mortise and tenon;
(D) Barefaced mortise and tenon; (E) Open-slot mortise with
key; (F) Fox-wedged tenon.

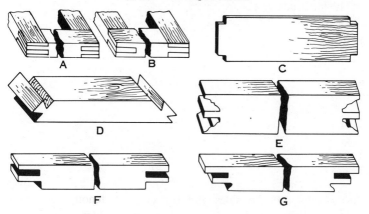

Fig. 5. Examples of some of the cuts possible on more elaborate
tenoning machines. (A) Double-joint tenon; (B) Single-joint tenon;
(C) Four-corner notch; (D) Dovetail corner block; (E) Door rail;
(F) Upper-sash check rails; (G) Lower-sash check rails.

231

SUMMARY

A mortise-and-tenon joint has always been recognized as one of the best methods of joining the end of one piece of wood to the side of another. Tenons can be made on circular saws, band saws, or on special tenoning machines, but mortises can only be made on mortising machines. The most common type machine used to make mortise joints in woodworking shops is the hollow-chisel machine. The hollow chisel has four sides of equal width, and is forced into the wood by the bit.

There are several types of tenoners, such as reciprocating plane, circular saw, rotary cutter, and hollow auger. Various machines have circular saws working at right angles to each other. One pair of saws running parallel to the grain of the wood cut the sides of the tenon, and another pair running at right angles cut the shoulders.

REVIEW QUESTIONS

1. What is a mortise-and-tenon joint?
2. How are mortises made?
3. Name four types of tenon machines.

CHAPTER 16

Sanding Machines

Abrasive tools play an important role in the woodworking industry. Chief among these tools are sanders, which are made in several different forms that include drum sanders, belt sanders, disk sanders, and spindle sanders.

The function of these machines is to smooth the wood surfaces by a scratching action of various types of abrasives. These abrasives come in a wide variety of sizes, and it is general practice with any given wood to use the coarsest grit possible that will not produce scratches visible to the eye.

Natural abrasives are found ready made in the earth and include sandstone, emery, flint, garnet, etc. Each has its own particular use. Flint is the least expensive and is the type of abrasive commonly found in sandpaper. Garnet, being much harder and tougher than flint, is an abrasive most commonly employed in the various types of sanding machines. Abrasive grains glued to sheets of cloth or paper are known as coated abrasives. Disks, sheets, drums, and belts are common examples of coated abrasives.

DRUM SANDERS

Drum sanders are generally large production machines used to sand large items, such as doors, plywood panels, and other types of flat work. The drum sander is also frequently used to surface the gluing faces of panel frames (after assembly of the frames on jigs) in order to smooth out any slight differences in adjoining frame members which might cause gaps in the glue bond.

The drum sander is built on the principle of a planer but, in place of the cutter heads, has one or more rotating sanding drums.

Certain types of drum sanders are designed to finish both the top and bottom surfaces of a board simultaneously and thus are equipped with sanding drums above and below the bed. The width of drum sanders vary for different applications and may be from two to eight feet.

The sanding drums consist of cylinders covered with a resilient material such as rubber or felt. In modern machines, each drum is driven by an individual motor and, as each drum rotates, it oscillates slightly from side to side, thus preventing the work from becoming scratched.

BELT SANDERS

Belt sanders are adapted to smoothing the planed surfaces of solid stock and plywood. Essentially, the belt sander consists of two or more pulleys supporting an endless abrasive belt, the pulleys being mounted on a cast-iron column. Between the columns is a table which can be moved at right angles to the belt. Either the table or the pulleys can be adjusted to the proper height.

There are four types of production belt sanders—the hand-block belt sander, the automatic-stroke belt sander, the hand-lever stroke belt sander, and the variety sander.

Hand-Block Belt Sanders

For ordinary sanding, belts about 6 inches wide are generally most practical for use on a hand-block belt sander. Paper belts are readily manipulated over the planer curved surfaces. For mouldings and irregular shapes, flexible, cloth-backed belts are used, ripped to whatever width best accommodates the members being sanded. When the standards are placed 12 to 20 feet apart, the sanding belts attain a high degree of flexibility which enables the operator to control the sanding of intricate shapes, and also increases his output. A typical single-belt hand-block sander is shown in Fig. 1.

In operation, the abrasive belt is brought into contact with the stock by applying pressure to a hand block or pad. The work is always in plain sight of the operator, providing sensitive control of the sanding, thus assuring accurate results.

Blocks for flat sanding are made of hardwood with the bottoms

Fig. 1. An open-belt hand-block sander with overhead idlers is one of the most widely used types of belt sanders.

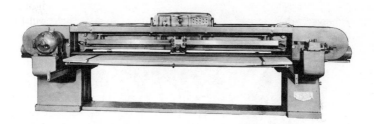

Fig. 2. An automatic stroke sander. A hydraulic system operates the sanding shoe. The work table is adjusted vertically by an electric motor to accommodate materials of various thicknesses, and moves back and forth under the sanding belt on roller bearings.

padded with felt and covered with a frictionless material to prevent wear. For rough sanding on plain surfaces, the hand blocks may be weighted by boring holes in them and filling with lead or babbit. There is practically no limit to the variety of ways in which hand blocks and pads may be adapted for sanding irregular shapes and special jobs.

235

Courtesy Mattison Machine Works

Fig. 3. A double-belt hand-lever stroke sander. This model features independent reversible motors on each of the sanding belts. One belt has a coarse grit and is used to remove tape and glue from veneered panels. The fine-grit belt is used to provide a final finish.

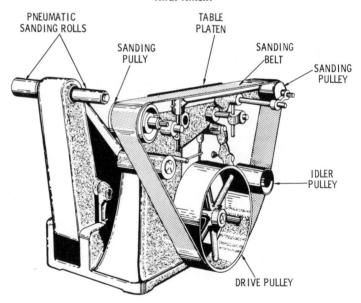

Fig. 4. A typical variety belt sander for rapid sanding of small and irregular-shaped objects.

Automatic Stroke Belt Sanders

The automatic stroke belt sander (Fig. 2) differs from the hand-block type in that the sanding block travels back and forth on a

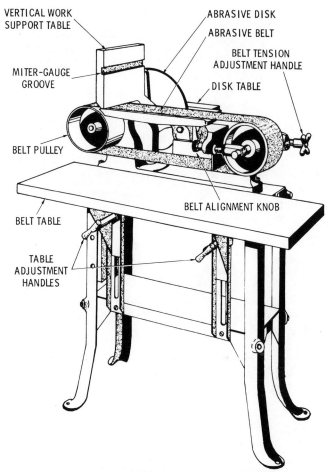

VERTICAL WORK
SUPPORT TABLE

ABRASIVE DISK

ABRASIVE BELT

BELT TENSION
ADJUSTMENT HANDLE

MITER-GAUGE
GROOVE

DISK TABLE

BELT PULLEY

BELT ALIGNMENT KNOB

BELT TABLE

TABLE
ADJUSTMENT
HANDLES

Fig. 5. A belt and disk sander. This type is ideal for the home workshop for fine cabinet and finishing work.

metal bar fastened between the posts. In this way, a uniform pressure (which can be regulated by the operator) is applied to the belt at all points.

Hand-Lever Stroke Belt Sanders

The hand-lever stroke belt sander (Fig. 3) also has a sanding block which travels on a metal bar, but the stroke of the block and

237

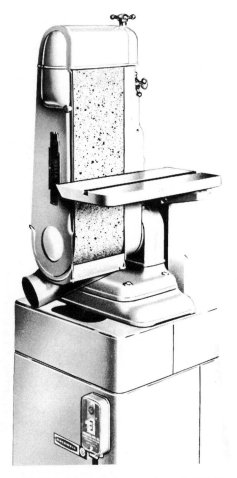

Courtesy Rockwell Manufacturing Company, Power Tool Division
Fig. 6. A typical belt sander.

the pressure is regulated by the operator by means of a lever fastened to the block. The work table, which supports the piece as it is sanded, is moved back and forth under the sand belt and, since it operates on ball-bearing rollers, very little effort is required.

Variety Sanders

The variety sander is especially useful for the rapid sanding

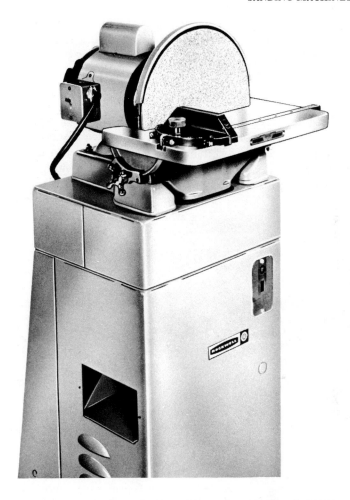

Fig. 7. A typical disk sander.

of small and irregular-shaped work which can be held free-hand against the open belt. Its belt runs over two or more pulleys of various sizes on which curved work of various diameters may be sanded. Some sanders of this type are provided with a means for attaching special forms and for various set-ups to accommodate different classes of sanding. Most variety sanders can be changed

239

from a vertical to a horizontal position, and some are equipped with a fence and a table which can be adjusted to various angles. One type of variety sander is shown in Fig. 4.

BELT AND DISK SANDERS

The combination belt and disk sander is used chiefly for cabinet work and other small stock, and consists principally of the parts shown in Fig. 5. The disk table may be moved up and down and tilted to any angle from 0° to 45°. It is also fitted with gauges for grinding compound angles, for making duplicate pieces, and for finishing circular pieces of various radii. The disk sander may be used for much circular work which is turned in the lathe, and also for many operations for which the planer and jointer are

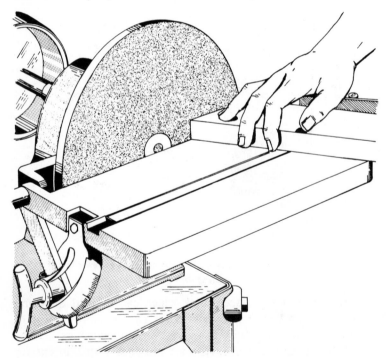

Fig. 8. A disk sander being used to sand the end grain of a piece of wood.

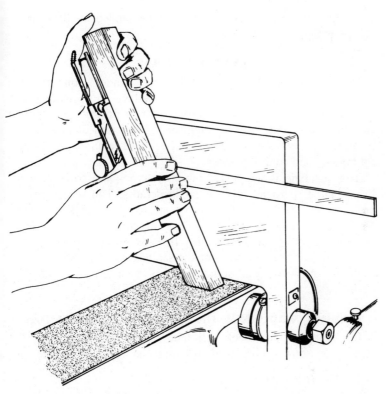

Fig. 9. *Use of the vertical table and miter gauge to sand the end of a piece of wood. The table assures an accurate surface that is 90° with relation to the sanding belt. The miter gauge determines the accuracy of the bevel.*

used. A separate belt sander is shown in Fig. 6 and a disk sander in Fig. 7.

The use of the miter gauge is shown in Fig. 8. By using this gauge in combination with the tilting table, it is comparatively simple to obtain any combination of angles desired on the work. Attachments are sometimes used in conjunction with disk sanders for grinding the teeth of gear patterns to the correct form. The speed of the disk at the rim should be about 7,000 feet per minute for ordinary work.

A miter table is also provided for the sanding belt. By using

241

this table and miter gauge, angles can be sanded on the belt, as shown in Fig. 9.

SPINDLE SANDERS

The spindle sander, also termed roll sander, consists of a substantial metal bench with a stationary level top provided with an adjustable throat opening through which projects the spindle carrying the sanding roll. Sanding rolls of different diameters may be attached to the spindle, which can be tilted to any desired angle for tapered and conical work.

This machine is adapted for both internal and external work having either perpendicular or inclined faces. The speed may be varied from 2,000 to 6.000 rpm by means of a regulating screw. The rolls are generally about 7 inches long and from 2 to 6 inches in diameter. The spindle carrying them is provided with an up-and-down movement to prevent the sanding material from cutting ridges in the work. The spindle may be sponge-rubber, or even air-inflated.

SANDING MOULDINGS ON THE HAND-BLOCK SANDER

In using the hand-block belt sander for finishing complicated mouldings, four important things must be carefully considered:

1. The right belt must be used.
2. The belt must be of the proper width for the particular moulding to be sanded.
3. The moulding must be supported in a fixed position on the table so the horizontal sanding belt will cover the profile of the moulding with the least twisting.
4. The hand block must be accurately shaped and carefully prepared to insure against dubbed corners.

Sanding belt manufacturers have developed a material for use with the hand-block moulding sanders and the edge-moulding and scroll-sanding attachments, which has great pliability as well as straight weave and minimum bulk, combined with fast cutting properties and long life. These characteristics are all very essential for a sanding belt used on mouldings having sharp corners or small beads. When buying sanding belts, be sure the character of the mouldings and the type of machine in use is taken into consideration.

In ripping sanding belts to width, allow about 3/4 inch more than the measurement of the profile of the moulding, so that the belt will ride easily in the slots of the hand block. Many mouldings have more members than can be covered by one width of belt. Also, some mouldings require a belt as narrow as 1 inch for reaching some members, and 3- or 4-inch belt for other members. In such cases, two or more operations are required.

After the belt is ripped to width and spliced, it should be carefully creased throughout its length at that point where it must follow a sharp angle in the moulding. This is accomplished (with the belt running on the machine) by holding a blunt instrument against a simple gauge which marks a straight line on the belt. Folding along the line (with the belt stopped) gives the desired crease, and the belt will follow the sharp corner without dubbing it.

The success of sanding mouldings of complicated profiles is due, in a large measure, to the care with which the hand blocks are made. Belts may be just right—of the correct width and flexibility and the mouldings may be properly supported under the belt. Still, unsatisfactory results will be obtained if the shape of the block does not conform exactly to the profile of the mouldings, or if provisions for protecting sharp corners are not made, or if slots (when required) are not carefully arranged.

In making blocks, select only well-seasoned hardwood (preferably maple), band saw it carefully to the profile of the moulding, then polish it to a perfectly smooth surface. Pieces of saw blade should be inserted wherever there is a sharp angle or corner (see Fig. 10) to protect the block and keep its corners from becoming rounded. Where a corner is to be kept sharp, it is necessary to provide a slot in the block for the edge of the belt to ride in. This slot can be made as indicated in Fig. 10. The side of the slot next to the abrasive side of the belt should be lined with a piece of steel to prevent the belt cutting the block.

Sanding mouldings with a hand block should offer no special problem. If care is used in making up the blocks, and a flexible

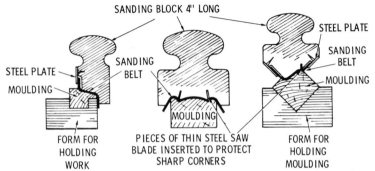

Fig. 10. Construction features of sanding blocks and forms for sanding mouldings.

light-weight sanding is selected, a little practice will soon enable the operator to produce perfect results.

SUMMARY

The function of a sanding machine is to smooth the wood surface by a scratching action of various types of abrasives. These abrasives come in a wide variety of sizes, shapes, and coarseness.

There are various types of sanding machines, such as drum, belt, disk, spindle, and vibrating. Drum sanders are generally large production machines used to sand large items, such as doors, plywood panels, and other large flat work. Certain type drum sanders are designed to sand or finish both top and bottom surfaces of a board simultaneously.

A combination belt and disk sander is used mainly for cabinet work and other small jobs. The disk sander is used for circular work and various angles. Generally, the sanding table can be moved up and down or tilted to any angle up to 45°. Gauges are also used to grind or sand compound angles, or for making duplicate pieces.

REVIEW QUESTIONS

1. Name the various types of sanding machines.
2. Name the type sanding machine used for cabinet work or for various angles.
3. What is a spindle sander?

CHAPTER 17

Boring Machines

Boring machines are made in various types and often in combination with other specialty woodworking devices. The most common type of boring machine is the single vertical-spindle type, fed either by a hand lever or a foot pedal.

Essentially, a boring machine consists of a vertical spindle having a chuck at one end and a telescoping splined sleeve with a drive pulley on the other. The spindle assembly is arranged to rotate in bearings attached to the frame of the machine. A table, usually capable of angular and vertical adjustment, is placed under or in front of the chuck to hold the work. Fig. 1 shows one type of boring machine, more popularly known as a drill press, suitable for light production use. A heavier machine for boring single holes is shown in Fig. 2.

In operation, a bit held in the chuck is forced into the work by hand feed. Some large woodworking mills are equipped with multispindle power-feed machines (Fig. 3) capable of boring a series of holes of the desired spacing, angle, depth, and size along the stock in one operation. Most boring machines are furnished with interchangeable spindles, thus adapting the machine for a wide variety of work. A key-type chuck used to hold straight-shank drills is shown in Fig. 4.

The size of a boring machine depends upon the size of work to be bored, and is hence figured according to the largest diameter stock which can be held on the table. Thus, for example, if the distance from the center of the table to the column is 10 inches, a 20-inch disk can be bored through its center, and the size of the boring machine is therefore 20 inches.

The speed range of a boring machine depends upon the size and number of pulleys, as well as the speed of the driving motor or shaft. In a typical machine equipped with a four-step cone

Courtesy Rockwell Manufacturing Company, Power Tool Division

Fig. 1. A typical drill press often found in home workshops and small cabinet shops.

pulley, the speed may be varied from 600 to 5,000 rpm. The speed of a machine used exclusively for metal working ranges between 400 and 2,000 rpm.

BORING TOOLS

All cutting tools used for making holes in wood are called *bits,* whereas similar tools used for metals are called *drills.* A distinction is sometimes made in the cutting operation, the term *boring* applying to holes, made in wood, while *drilling* means a cutting operation in metal. The most common type of machine boring bits are the *auger* and *twist-drill*. Other types are the *center*

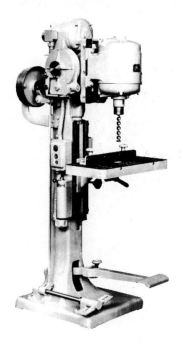

Courtesy B. M. Root Company

Fig. 2. A heavy-duty boring machine designed for continuous production use.

and *expansive bit,* each of which has its particular application in the woodworking shop.

Auger bits (Fig. 5A) are equipped with a central screw which draws it into the wood, two ribs which score the circle, and two lips or cutters which cut the shavings. These are used for boring holes from 1/2 to 2 inches. The sizes are listed in 16ths, thus a 2-inch auger is listed as a number 32.

A twist drill differs from an auger in the absence of a screw and a less acute angle of the lip, hence there is no tendency to split the wood. In other words, the tool does not pull itself in by

247

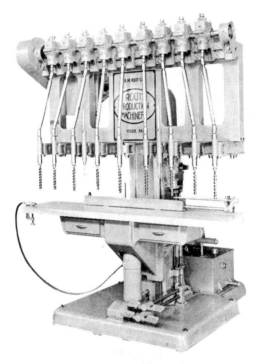

Courtesy B. M. Root Company

Fig. 3. A multispindle boring machine capable of boring a series of holes at the desired angles, spacing, and depth, all in one operation.

a taper screw but enters by external pressure. Twist drills are used for boring small holes where the ordinary auger would probably split the wood. They come with either round or square shanks and in sizes from 1/16 to 5/8 inch or larger, varying by 32nds.

The center (Fig. 5B) and expansive bits (Fig. 5D) are used for boring large holes. The center bit is used primarily for boring holes through material which might split with bits of other types, whereas the expansive bit is adjustable and is used for boring large holes of various diameters, usually up to 3 inches. Another type of boring device is the Forstner bit (Fig. 5C).

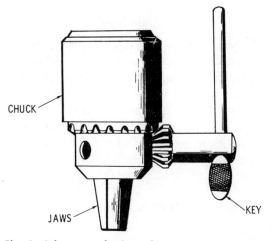

Fig. 4. A key-type chuck used in a typical drill press.

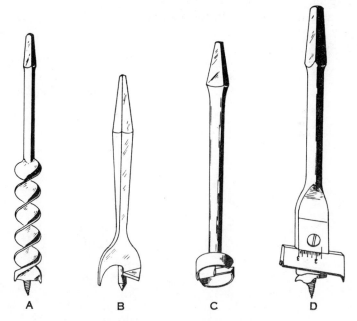

Fig. 5. Various types of bits for boring holes in wood; (A) Auger bit; (B) Center bit; (C) Forstner bit; (D) Expansion bit.

249

SUMMARY

Essentially a boring machine, more popularly known as a drill press, consists of a vertical spindle having a chuck at one end and a telescoping splined sleeve with a drive pulley on the other. A table is placed under or in front of the chuck to hold the work.

The size of the boring machine depends on the size of work to be bored and the speed of the drill. The cutting tools used for making holes in wood are called bits, whereas similar tools used to bore metal are called drills.

REVIEW QUESTIONS

1. What is a drill?
2. What is a bit?
3. Explain the purpose of a drill chuck.
4. What is the difference between an auger bit and an expansion bit?

CHAPTER 18

Power Operated
Hand Tools

Power operated hand tools, as distinguished from woodworking machinery, are used on actual building construction, and for the most part, take the place of hand tools. It is frequently impossible, or at least not practical, to install heavy, automatic woodworking machinery for this type of work, so recourse to more readily portable power-operated tools is made. Where power (electricity or compressed air) is available, many types of power tools are now used. Among these are electric hand saws, hand planes, routers and mortisers, various types of air and electric hammers (especially suited for drilling masonry and concrete), heavier jack hammers (used to break up concrete paving), and various types of nailing and stapling machines. The electric floor-sanding machine has taken the drudgery out of one of the most disagreeable of the carpenter's jobs. On the large carpentry jobs, an electric bench grinder is useful, and power-operated saw filers and grinders to sharpen circular saws are often found.

POWER HAND SAWS

These types of saws are manufactured in various sizes and are commonly driven by an electric motor, with the entire mechanism enclosed, as illustrated in Fig. 1. These saws are very easy to operate and are made to cut material up to approximately four inches thick, depending upon the size of the saw. In making a cut with a saw of this type, proceed as follows:

251

Fig. 1. An electric hand saw of the type used by many carpenters.

With the switch in the off position, rest the front edge of the saw base flat on the work. Start the saw by pulling back on the trigger switch and begin the cut, being careful not to jam the saw blade into the work suddenly. Forcing may place an unnecessary strain on the operating parts, which might possibly require their premature replacement. For most cutting operations, guiding the saw through the work is all that is necessary. If the motor should stall due to a dull blade or unnecessary pressure, do not release the switch at once, but pull the saw back, allowing the blade to run free before shutting off the motor. This precaution will reduce burning of the contact points in the switch and greatly lengthen its life.

Power Source

Before connecting the cord plug to an electrical outlet, be sure that the line voltage is the same as that stamped on the name

252

plate of the tool. When an extension cord is necessary, it should be of sufficiently heavy wire to assure full voltage at the tool with the machine under load. As a safety precaution, and particularly in a wet or damp location, the ground wire projecting from the side of the connector should be attached to a suitable ground, such as a water pipe, or to a permanently grounded conductor.

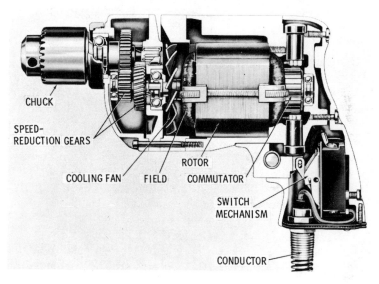

CHUCK

SPEED-
REDUCTION GEARS

COOLING FAN   FIELD

ROTOR

COMMUTATOR

SWITCH
MECHANISM

CONDUCTOR

Fig. 2. Cutaway view of a typical portable electric drill.

Saw Blades

Although saws are most frequently equipped with combination blades for general-purpose work, special blades for ripping, cross-cutting, mitering, dadoing, metal cutting, etc., are available. Be sure when changing blades that the blade teeth are pointing in the direction of rotation of the saw. Abrasive disks for cutting asbestos-cement boards and concrete are also available.

ELECTRIC DRILLS

The process of drilling holes in metal and boring holes in wood with an electric drill is similar to drilling or boring by hand, except

253

that the power for turning is furnished by an electric motor instead of by the operator. Drills of this type usually have capacities for drilling holes from 1/16″ up to 1″ in diameter.

Fig. 2 shows a cutaway view of a popular type of electric drill equipped with a pistol grip or spade handles, and with a geared chuck which automatically centers the drill shank in the tool. Many electric drills can be fitted with attachments for driving screws, rotating small grinding wheels, drilling at right angles, etc.

Operation

In drilling metal with an electric drill, the operator must first be sure that the diameter of the hole to be drilled is within the capacity of the tool. Electric-drill sizes usually indicate the largest diameter drill the tool will accept. For example, a 1/2-inch electric drill is intended to drill holes up to and including 1/2″ in diameter, and no larger.

It is better practice, where possible, to use a drill which has the capacity to drill somewhat larger holes than the work calls for. The location of the hole should be carefully marked and started with a center punch, exactly as when drilling by hand. Then, with the motor running, insert the point of the drill into the punch mark and start drilling. Care must be used to hold the electric drill at right angles to the work so the hole will be straight. With the tool held in this manner, exert a light pressure and continue drilling.

If the hole is to go completely through the work, relieve the pressure on the drill when the point of the drill bit begins to break through and until the hole is completed. Finally, withdraw the drill from the hole by pulling it straight back, and then shut off the motor.

Twist drills do not pull themselves into the work; they must be fed by pressure, and this pressure must be exerted by the operator of an electric drill in exactly the same way as if drilling entirely by hand. The only effort saved the operator by the electric drill is that of turning.

Drill Stand

A drill stand aids in accurately locating and maintaining the direction of a hole to be drilled, as well as providing the operator

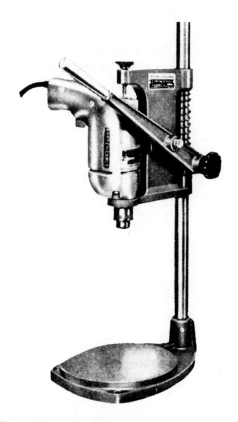

Fig. 3. A drill stand can be used to mount an electric drill.

with an easy control for feeding the drill into the work. A common type of drill stand is shown in Fig. 3 with an electric drill fitted to it. A lever is provided on such a stand so that the operator can feed the drill into the work with either very heavy or comparatively light pressure.

When a drill stand is used, the work is placed on the table provided and the tool brought down on it by means of the handle. The location of the hole, therefore, must be placed under the drill, rather than the drill put to the location, as when drilling by hand. The work should be securely fastened to the table with clamps, and the drill fed into it by means of the lever with sufficient pres-

255

sure to cut, but not enough to cause the drill to overheat or the motor to stall.

For accurate work, a drill stand should be used if one is available. The stand is so constructed that the hole will always be accurately drilled at right angles to the work's surface, or at any other angle if the work is clamped to the table accordingly.

BENCH GRINDERS

Bench grinders are commonly used in woodworking shops for sharpening chisels, screw drivers, etc., and for smoothing metal surfaces, etc. Fig. 4 shows a common type of bench grinder. This type of grinder consists mainly of an electric motor having a double-ended horizontal spindle, the ends of which are threaded and fitted with flanges to take the grinding wheels. Other models employ a conventional belt drive.

The size of the grinder is commonly taken from the diameter of the abrasive wheel used in connection with it. Thus, a grinder with

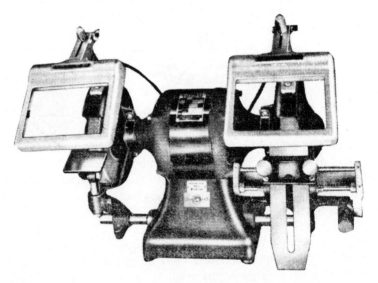

Fig. 4. A bench grinder used to sharpen tools.

a 6-inch diameter wheel is called a 6-inch grinder. Grinder units are further classified as bench or pedestal, the latter indicating a floor model.

A bench grinder is usually fitted with both a medium-grain and fine-grain abrasive wheel. The medium wheel is satisfactory for rough grinding where a considerable quantity of metal has to be removed, or where a smooth finish is not important. For sharpening tools or grinding to close limits of size, the fine wheel should be used, as it removes metal slower, gives the work a smoother finish, and does not generate enough heat to anneal the cutting edges. When grinding tools, keep a pan of water handy, and dip the tool in it often to assure against overheating.

When a deep cut is to be taken on work, or when a considerable quantity of metal must be removed, it is often practical to grind with the medium wheel first and finish up with the fine wheel. The wheels are removable, and most bench grinders are so made that wire brushes, polishing wheels, or buffing wheels can be substituted for the grinding wheels.

Operation

In grinding, the work should be held firmly at the correct angle on the rests provided and fed into the wheel with enough pressure to remove the desired amount of metal without generating too much heat. The rests are removable, if necessary, for grinding odd-shaped or large work. As a rule, it is not advisable to grind work requiring heavy pressure on the side of the wheel, as the pressure may crack the wheel. As abrasive wheels become worn, their surface speed decreases and this reduces their cutting efficiency. When a wheel becomes worn in this manner, it should be discarded and a new one installed on the grinder.

Safety Precautions

Before using a bench grinder, make sure that the wheels are firmly held on the spindles by the flange nuts, and that the work rests are tight. Wear goggles, even if eye shields are attached to the grinder, and bear in mind that it is unsafe ever to use a grinder without wheel guards. Wearing glasses does not take the place of wearing goggles. A pair of expensive glasses is easily ruined, as the flying sparks stick to the glass, and sometimes cannot be polished off. Also, remember that it is easy to run a thumb or finger into the wheel.

ELECTRIC PLANES

In finish carpentry work, the use of a plane is often necessary. For example, in fitting doors, it is usually necessary to plane one or more edges to obtain the proper clearance between the door and its jamb. A plane is also a necessity when cabinets are built on the job, and for the installation of various other items, such as book shelves.

Electric planes are now available to perform much of the work formerly done with hand tools. An electric plane, such as the one shown in Fig. 5, provides an accurate and rapid means of planing and, in most instances, will result in more precise work with less effort. The model shown is adjustable for depth of cut and has a side fence that can be set to plane any desired angle. When set at 90°, the planed edge will be at a true right angle to

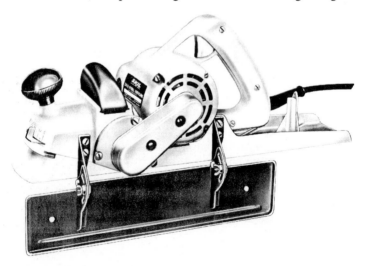

Courtesy Skil Corporation

Fig. 5. An electric plane. This model is light-weight and has a 3" cut.

the side of the material. The plane in Fig. 5 has a 3-inch cutter which is more than adequate for most work.

SABER SAWS

Another relatively recent addition to the carpenter's tool collection is the electric saber saw (Fig. 6). The carpenter will find this one of the most versatile tools he owns. Most models are capable of cutting through 2-inch stock with ease and can be used in places impossible to reach with an electric handsaw. Intricate shapes can be cut in plastic, wood, and even metal with this saw by using the proper type of blade. The cut can be adjusted to any desired depth and angle.

A full-range of blades is available for this type of saw, from the standard combination blade for rough cuts in wood, to metal-cutting blades. Even knife-edge blades for cutting leather, fabric, etc., may be purchased. Accessories are also available. These include fences for accurate straight-line cuts, circle-cutting attach-

Courtesy Skil Corporation

Fig. 6. A two-speed heavy-duty jig saw.

ments, and offset blade chucks to permit sawing flush with a wall or sawing up to an object.

PORTABLE SANDERS

A tool that eliminates much of the drudgery from finishing and, at the same time, produces a better finsh is the portable electric sander. There are many types of these handy machines available—the orbital sander, reciprocating sander, and the belt sander. Of the three, the belt sander, such as the one in Fig. 7, provides the most rapid removal of material. The orbital or reciprocating sanders are better suited for final finishing work, however. All grades of grit are available for the belts and sheets used with these machines. The sander shown has a belt which is 4 1/2″ wide.

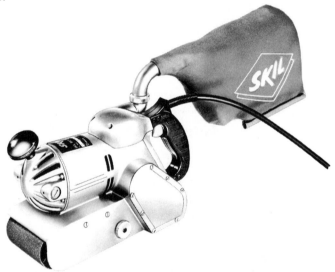

Courtesy Skil Corporation

Fig. 7. A 4½″ belt sander with a dust catcher.

SUMMARY

Power-operated hand tools include electric hand saws, hand

planes, routers and mortisers, air and electric hammers, portable drills, bench grinders, and hand sanders. Many of these tools have taken the disagreeable jobs out of carpentry work.

Power hand saws are very easy to operate and are made to cut material up to approximately 4 inches thick, depending on the size of the saw. It should be noted that, although very easy to operate, this type of power tool can be dangerous in operation due to suddenly jamming the saw blade into the wood. Generally, saws are equipped with combination blades for general-purpose work, but special blades for ripping, crosscutting, mitering, dadoing, metal cutting, etc., are available.

Drilling holes in metal or boring holes in wood with an electric drill is similar to drilling or boring by hand, except that power for turning is furnished by an electric motor instead of by the operator. Many drills can be fitted with attachments for driving screws, rotating small grinders, drilling at right angles, etc. Many drills can be mounted on a drill stand which provides an easy control for feeding the drill into the work.

REVIEW QUESTIONS

1. Name a few power-operated hand tools.
2. What are some advantages in using a saber saw?
3. Why are various type saw blades manufactured for the power hand saw?
4. Name the three types of portable hand sanders.
5. What are some advantages in using an electric plane?

CHAPTER 19

Termite Protection

At certain times of the year, termites develop wings, emerge from the ground in swarms sometimes like bees, and fly away to form new colonies. Winged termites and winged ants look somewhat alike, but they have definite differences, as seen in Fig. 1. Termites have a straight body and four milky-white wings, all of equal length and twice as long as the body. Swarming ants also

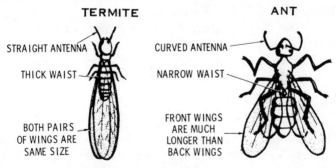

Fig. 1. Termites and ants differ in several respects.

have four wings, but they are transparent and of unequal length, the longest only half again as long as the body. Termites have short, straight, beaded antennae, while the antennae of ants are elbowed, with a bead-like end.

Termites prefer to enter a building through cracks in the foundations, or through the cavities in a hollow wall. When these avenues are not available, they will build mud tunnels up from the ground to the wooden parts of the building. These tunnels will be on the surface of the walls (usually on the dark interior side), or back of a pile of lumber, brick, or other materials stacked against the foundation. They are great engineers, however, and have often been known to build hollow tubes up from the ground for a foot or more to reach edible wood. Porch floors consisting of a concrete slab over a fill are fruitful sources of termite troubles, and exceedingly hard to eradicate. It may necessitate drilling through the floor and injecting heavy doses of toxic chemicals. It may be necessary to do the same where concrete drives or walks contact foundations.

In the worst cases of termite infestation, it is probably best to employ a competent pest-control firm. Their work is usually guaranteed, and they are often bonded.

It has been estimated that during the year 1963, wood-eating termites destroyed $250 million worth of property in the U.S., or more than the combined loss from arson, tornadoes, and lightning, and more than twice the damage they caused only ten years ago. Measures to assure their control are becoming more and more necessary. Some of the methods used have not proved to be entirely satisfactory. Certain woods are susceptible to *Lyctus* (powder-post) beetles. If a powdery substance is noticed coming from small holes, these beetles have attacked the wood.

TERMITES AND THEIR IDENTIFICATION

There are several kinds of termites, but the ones who do the most damage are of the subterranean species. This type lives in the ground, but travel into wood which is in contact with the soil, or into wood which is close enough to the soil to permit building mud tunnels through which they can reach the wood. The presence of these tunnels is sometimes the first sign of a termite infestation. Only a few species of timber are immune to termite damage. Heart cypress and redwood have some slight immunity, but even these woods are occasionally attacked. They do not seem to bother very resinous heart wood of long-leaf southern pine, but this kind of timber is not always available. Pressure-treated with creosote, all species of timbers are entirely immune to termite attack.

Termites are often found in stumps or posts and other wood in contact with the soil, but even though such infestations is very near a house, it does not imply that the house is, or will be, infested. Termites seldom leave a location of their own accord, and they work very slowly. There is no need to panic when they are discovered, for the damage, to a great extent, is already done. A delay of even several months is usually of little consequence.

CONSTRUCTION TO PREVENT TERMITE DAMAGE

Termites must maintain contact with the ground to obtain the moisture necessary for their existence. Hence, the first consideration should be to build in a manner which will prevent the entry of termites from the ground. The foundations of buildings should be constructed either of masonry or of approved pressure-treated or naturally termite-resistant lumber. Where a basement is provided, the foundation walls should be of masonry. If unit-block construction is employed, such as brick, tile, cement blocks, etc., all joints should be well filled with mortar and the wall topped with a 4" cap of concrete. This should be reinforced to prevent cracking when over open-type units. The ground within the basement should be sealed over with concrete; posts should not extend through the floor into the soil, but should rest on concrete footings that extend at least 2" above the floor.

If foundations are built over an earth fill or naturally loose earth, subsequent setttlement may cause the joints between the concrete basement floor and foundation walls to open up. Such joints are a probable source of termite entry and should be guarded against by installing mastic or metal expansion joints between the walls and floor. Concrete for basement floors and walls should be a dense mixture and the walls reinforced with steel rods at the corners and intersections to tie them together. Improper construction of the basement floor and walls may cause cracks to develop through which termites may gain free access from the earth. Window sills and frames in the basement should not come in direct contact with the ground, nor should leaves or debris be allowed to collect and remain in contact with them.

Buildings which have no basements should have the sills set a minimum of 18″ and preferably 24″ above the excavated ground or natural grade at all points to afford the necessary clearance and good ventilation. This is shown in Fig. 2. On the exterior, the building clearance to woodwork may be reduced to 8″ above the finished grade line, provided the foundation walls permit access to occasional inspection for shelter tubes by the home owner. In the case of solid foundations, ventilation should be provided by allowing not less than 2 square feet of net open area for every 25 linear feet of wall. Openings should be screened with 20-mesh noncorroding screening. All lumber which comes in contact with the ground should be pressure treated with a preservative.

TERMITE SHIELDS

As previously pointed out, termites require constant access to

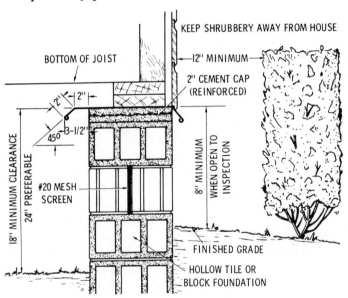

Fig. 2. *Metal shields used in various types of construction to protect against an invasion of termites.*

soil moisture and, unless they stay in a moist atmosphere, they soon die. Because of this need for moisture, termites construct shelter tubes of earth or waste material to carry moisture and to act as passageways between the ground and their food supply when it is not in contact with the soil. Destroy or prevent this ground contact and the termites cannot damage the building.

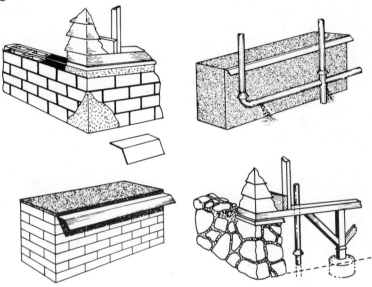

Fig. 3. A termite shield on top of a hollow foundation wall in a building with no basement.

Termites may build these shelter tubes over the face of stone, concrete, brick, or timber foundations, and along water pipes or similar structures. Such contacts may be prevented by means of a metal shield barrier. This termite shield consists of noncorroding metal firmly inserted and pointed into a masonry joint or under the sill. It projects horizontally at least 2″ beyond the face of the wall and then turned downward an additional 2″ at an angle of 45°. All joints should be locked (and preferably soldered also), with the corners made tight and the outer edge rolled or crimped to give stiffness against bending as well as to eliminate a sharp edge.

The termite shield should be used on each face of all foundation walls, except that it may be omitted from a face (either interior or exterior) which is exposed and open to easy and ready inspection. However, around houses and places where srubbery may partly conceal the wall and inspection is likely to be infrequent, a modified shield may be used. Here, the horizontal projection is omitted and the 2″ projection bent downward at 45° is employed. Metal termite shields are required by many codes and by some city ordinances. Fig. 3 shows some typical termite

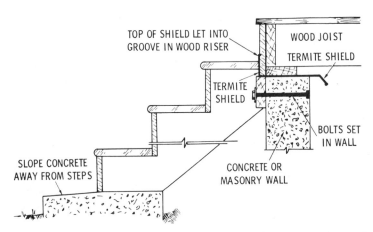

Fig. 4. Location of a termite shield between wooden steps and a porch.

shields. A method of inserting shields to prevent the passage of termites by the way of wooden porch steps is shown in Fig. 4.

DETECTING AND LOCATING TERMITE DAMAGE

Except at swarming time, termites are hidden unless their galleries or passageways are broken open. The earth-colored shelter tubes offer a ready means of recognizing their activities. Their work can usually be detected in wood by striking it with a hammer. Solid wood rings clear, while timber or woodwork eaten out by termites will give a dull thud when struck. Striking with an ice pick also determines weakened wood.

Clean-cut holes in books, papers, and clothing are good indications of the presence of termites. Springy basement floors or the softening or weakening of woodwork suggests termite, or other damage.

STOPPING TERMITE DAMAGE

The most lasting and effective remedy for termite damage is to replace any wood in or near the basement of the building with concrete. Second in order of effectiveness and durability is to replace such wood with treated wood or timbers and, in regions of excessive termite damage, to employ protective termite shields. By this means, contact between the colony and the building is permanently broken and relief from termite damage is assured. This means that joists imbedded in concrete, wooden basement floors, and baseboards should be replaced with any type of plain or ornamental concrete. In basement rooms so constructed, movable furniture of wood, and also built-in furniture, particularly if resting on concrete footings, can be employed with safety.

To cap and face basement walls of frame buildings, it is rarely necessary to jack up the building. It is usually possible to remove the upper tier of brick or upper portion of the masonry unit in sections and replace with Portland-cement mortar and a suitable cap of slate or mortar. Where poor grades of mortar have been used in masonry walls below the ground, it may be necessary to coat the outside and, if necessary, the inside of the wall with concrete to keep termites from boring through.

USE OF SOIL POISONS

The use of soil poisons is one of the most effective means of termite control. Its success lies in preparing a complete chemical barrier which leaves no point of entrance for the termites. Two of the most effective and least expensive of the preparations used for this purpose are *chlordane* and *dieldrin*. Both are toxic, but not so much so that they are dangerous to handle. Dieldrin is the most highly toxic, equal in toxicity to lead arsenate or straight arsenic. Chlordane has a toxicity about equal to Bordeaux mixture, lime-sulphur sprays, DDT, or aspirin. To use dieldrin, mix one

269

gallon of 15% emulsifiable dieldrin concentrate with 44 gallons of water. To use chlordane, mix one gallon of 45% emulsifiable concentrate with 44 gallons of water.

To apply, dig a V-shaped trench against the foundation walls (both sides, if possible) and loosen the soil deeper if possible, so the chemical will penetrate to the bottom of the footings. Pour 2 gallons of the chemical emulsion into each 5 linear feet of the trench. Pour 4 gallons along each 5 linear feet of basement walls. Neither of these chemicals will injure shrubbery or other growing plants.

It is well to inspect the work about 3 to 4 weeks after the poisoning is completed, and re-poison if there is any indications of continued termite activity. The building should be regularly inspected about 4 weeks after the semi-annual "flying time" and re-poisoning done if there is evidence of new infection.

SUMMARY

There are several types of termites, but the ones who do the most damage are of the subterranean species. This type lives in the ground and travels to wood through mud tunnels. The presence of these tunnels is sometimes the first sign of termite infestation. Heart cypress and redwood have a slight immunity, but even these woods are occasionally attacked.

Termites must maintain contact with the ground in order to survive. The first consideration should be to build in a manner which will prevent the entry of termites from the ground. A foundation of any building should be constructed either of masonry or of approved pressure-treated or naturally termite-resistant lumber.

Where a basement is provided, the foundation walls should be of masonry. If construction of basement walls is of block, all joints should be filled with mortar and the wall topped with a 4-inch cap of concrete.

A termite shield may be installed under the foundation sill to prevent termites from entering the wood. This is a noncorroding metal shield which projects horizontally at least 2 inches beyond the foundation wall and then is turned down at a 45° angle an additional 2 inches. All joints should be soldered together and corners made tight with the outer edge crimped to give stiffness.

The use of soil poisons is another effective means of termite control. Two chemicals most commonly used are chlordane and dieldrin which are poured into V-shaped trenches against the foundation walls.

REVIEW QUESTIONS

1. Name several ways to prevent termite damage.
2. What are two chemicals used in the ground to prevent termites from entering?
3. What should be done to the top row of blocks in a foundation to prevent termite damage?
4. Name two woods that have a slight immunity to termites.
5. What connection must a termite have with respect to ground?

CHAPTER 20

Painting and Equipment

Paint is *a mixture consisting of finely divided substance, called pigments, held in suspension in a liquid called the vehicle.* In this condition it is capable of being spread on the surface by the application of a brush or sprayer. When the surface has been covered, it is said to have received a coat of paint. Usually two or three coats are applied, the second and third coats being applied after an interval of time sufficient for the preceding coat to dry. It should be noted that three coats of thin paint will last longer and less material used than two coats of heavy paint.

PAINTS

The pigments or different solids which are used in paints do not become paint until it is mixed with the thinning material or vehicle. Paint may be obtained in two forms—unmixed or paste, and mixed. The so-called unmixed paints consists of the pigment ground in just enough of the vehicle to form a paste; the painter adds to this more of the vehicle to properly thin the paint. The mixed paints are mixed by the manufacturer, and since a good vehicle is expensive, the buyer should purchase a well-known brand.

The base materials used in producing paints are:

1. Pigments.
2. White lead.

3. Sublimed white lead.
4. Zinc white.
5. Barytes.
6. Whiting.
7. Vehicle.
8. Linseed oil.

Pigment

Pigment is a powdered substance mixed with a suitable liquid, in which it is relatvely insoluble, to form paints. The principal white base pigments are white lead and zinc white. In addition to these are several other white pigments which are variously regarded as inert pigments or extenders. Among these are barytes, whiting, and gypsum.

White Lead

White lead is a heavy, white, poisonous powder consisting of a basic lead carbonate, usually having the composition $2PbCO_3$-$Pb(OH)_2$, and forming an important pigment or paste formed by grinding it with oil. In making white lead, pig lead is melted by means of a specially designed machine and subjected to a blast of superheated steam, which reduces the metal to its finest possible state of subdivision. This process partly hydrates and oxidizes the comminuted lead which passes into cylinders equipped with agitators and containing water through which a current of air is passed.

The subhydroxide of lead is further oxidized into lead hydroxide. It is then treated in other cylinders containing water with a stream of purified carbon dioxide gas which converts it into a basic white lead of ideal composition, after which the completed product goes to specially designed drying beds.

Sublimed White Lead

This is a lead sulphate that is white in color and is produced by a fire process. It is used in making ready-mixed paint, and is inferior to the genuine white lead.

Zinc White

Zinc white is an oxide used as a white pigment in house paints, water colors, etc. This is zinc made by blowing the current of air through molten zinc. It is free from poisonous qualities which render white lead objectionable for an indoor paint, and is not acted upon by sulphur or gases, which darken white lead. Zinc white spreads more rapidly than white lead, but does not cover so well. It requires four or more coats of zinc white to equal the covering obtained by three coats of white lead. It is generally used in combination with white lead on exterior paints.

Barytes

This is a sulphate of barium and is a native rock which is very finely ground and is used as an inert extender of paint. It has no affinity for linseed oil, and absolutely no covering power. Four coats of barytes mixed with linseed oil will not hide a surface which is being painted. The same is true of silica, which is another so-called inert pigment.

Whiting

This is a pulverized chalk, calcium carbonate, and is lighter in weight than braytes, and possesses a decided affinity for linseed oil. Mixed with oil in a paste form, it becomes putty. It has very little covering power in itself.

Vehicles

The best vehicles or thinning agents are linseed oil and turpentine. In mixing paint, a small proportion of one of the driers is added to these vehicles to make the paint dry in a reasonable length of time. In enamel paints, the vehicle is varnish, and in kalsomine and other cold-water paints, it is a solution of glue, casein, albumen, or some similar cementing material. The cementing material is sometimes called the binder. A vehicle for outside paint of the best quality will generally contain 90 to 95% linseed oil, and from 10 to 15% japan driers.

Linseed Oil

Linseed oil is the principal vehicle used in paint and is the oil expressed from the seed of the flax. It is a drying oil which gives it a great value in paints. Linseed oil is adulterated with various mineral oils, resin oil, corn oil, and fish oil. Linseed oil may be obtained either in raw or boiled forms. The object of boiling linseed oil is to drive off part of the moisture which the raw oil contains. After boiling, some form of the driers are added to hasten its drying. The result is a heavy-body partially oxidized oil.

When genuine boiled oil is used, the paint dries rapidly on the surface but the under part of the paint film remains soft and tacky for a long time. If the final coat is applied too soon, the pulling of the brush on the soft under coat will cause it to shrivel up and crack. In most cases, blistering will occur. Do not use boiled oil on any exterior work.

Turpentine

This vehicle is the distilled sap of the long-leaf pine. It is used to make paint more fluid, and to make it spread easier. It dries by evaporation, leaving a slight gummy residue. Benzine is a by-product of petroleum and usually costs less. It dries by evaporation like turpentine *but* it will not leave a residue.

Driers

In order to make the paint dry within a reasonable length of time, oxidizing agents or *driers* are mixed in the paint. They usually consist of salts of lead or salts of manganese. Some driers are also called japans. Not more than 10% by volume of any liquid drier should be added to oil. Excess of drier makes the paint less durable. Cheap driers often contain resin.

PAINT MIXTURES

There are three kinds of mixtures of color, which are oil, flatting, and distemper. The oil is bright and glossy, the *flatting* is perfectly flat or dead (without gloss), and the *distemper* is like the flatting except it is not as durable. The chief body in oil and flatting color is white lead, but in distemper or water-color, whiting is substituted. The three ordinary vehicles in mixing are oil for oil colors, turpentine for flatting, and water for distemper. In addition to these, potent driers are used in oil colors, japanner's gold-size and varnish in flatting, and glue size in distemper.

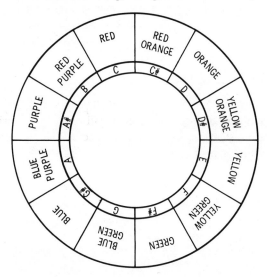

Fig. 1. Illustrating the three primary colors with the secondaries and tertiary colors shown.

PRIMARY COLORS

The diagram shown in Fig. 1 shows the three primary colors, their secondaries, and what may be called the tertiary colors. Opposite each of these there has been placed one of the notes of the chromatic music scale forming a perfect octave. It is interesting to note that the claim has been made, and with much insistence, that any scheme of color that may be selected and which may be struck as a chord will, if the chord is harmonious, become a harmonious scheme of color. If this chord produces a discord of music, there will be a discord of color.

By definition, primary colors are those that cannot be made by mixing two or more colors together. The three primary colors are, *red, blue,* and *yellow.* Fig. 2 shows a diagram of the primary and secondary colors. The colors obtained from mixing any two of the primary colors together are called secondary colors. There are three secondary colors—*purple, green* and *orange.*

277

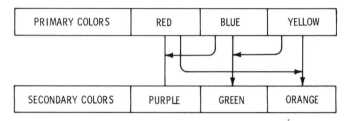

Fig. 2. A diagram of primary and secondary colors.

Red and blue gives *purple*
Blue and yellow gives *green*
Red and yellow gives *orange*

By mixing any two of the secondary colors together you get what are called tertiary colors, which are: *citrine, olive,* and *russet.*

Orange and green gives *citrine*
Green and purple gives *olive*
Orange and purple gives *russet*

Black and white are not regarded as colors. A good black can be produced by mixing the three primary colors together in proper proportions. By adding white to any color you produce a tint of that color. By adding black to any color you get a shade of that color. That is the difference between *tint* and *shade.* The use of black subdues or lowers the tone of any color to which it is added. To preserve the richness of colors when you wish to darken them, use the primary colors instead of black. To make a yellow darker, use red or blue, and to darken blue, add red and yellow, and so on.

Every shade or tint of color required by the painter can be made from red, blue, and yellow with black and white. To make any of the umbers or siennas lighter in color and to preserve the clear richness of tone, always use lemon chrome instead of white. If you wish a subdued or muddy umber or sienna color, then use white.

The most useful primary colors are:

278

Yellows—*lemon chrome, deep ocher*
Reds—*vermilion, Venetian red, crimson lake*
Blues—*Prussian blue, ultramarine*

Gold or silver leaf harmonizes with all colors and, with black and white in small quantities, can be used to bring into harmony the most glaring colors. The old time heraldic painters knew the value of outlining their strong primary colors with gold, silver, black, or white, to bring them into harmony with each other. The Egyptians and other ancient people made use of the same knowledge in their decorative schemes. Yellow ocher is the most useful color the painter possesses. In its pure state it is admirable for large wall spaces, and if you are in doubt as to what color to use to complete a color scheme, you will find ocher or one of its shades or tints will supply the missing link.

Red on walls make a room look smaller and will absorb light. Yellow gives light and airiness to any room and it will also reflect light. Useful colors in large quantities for churches, public halls, etc., are:

Primrose red
Terra cotta (white, burnt sienna, lemon chrome)
All tints of ocher
Flesh colors (white and burnt sienna)
Pea green, apple green
Grey green (white, paris green and a touch of black)
Ivory shades (white, lemon chrome, or ocher)
Old rose (white, ocher, Venetian red, or pure Indian red and black)
Nile blue and Nile green (white, Prussian blue, lemon chrome)
Light citrine, light olive, light russet

For ceilings the best tints are the creams and ivory tints, and grey. Creams and ivory tints are made from white tint with one or more of these colors.

Lemon chrome

Orange chrome
Ocher
Raw sienna

To produce a warm tone, add a small quantity of burnt sienna, vermilion, or Venetian red. To produce a colder tone, use a little green, black, raw umber, or blue. Greys are made from white tint with either black, black and green, blue and umber, black and red, red and blue, or burnt sienna and blue. Light colors are used for ceilings in preference to dark colors. Contrasting colors are better for ceilings than a lighter tint of the wall color.

INTERIOR PAINTING

Exterior paints or varnishes are seldom a good choice for interior surfaces because they may yellow with age and may be too slow in drying. Paint finishes must vary to meet the requirements for appearance and reflection of light indoors. High-gloss enamels or varnishes are too harsh in appearance and in light reflection to be suitable for surfaces of large areas, although they are often appropriate for wood trim and furniture. Some of the high-gloss coatings can be rubbed with pumice stone and oil to any desired finish, but the expensive labor required usually eliminates this operation. There are two classes of gloss enamels, one sufficiently rich in pigments for the enamel to be used for priming or undercoat as well as for finish coat, and the other requiring specially prepared material for the undercoating. The former is less expensive and is by far the more widely used, with two coats usually sufficient.

When the most elegant appearance is demanded, sanding of the wood beforehand and sanding the undercoat after it has dried is essential. At least two coats of undercoater and one of enamel are usually needed for good results. Semigloss enamels or interior paints are lustrous without being harsh. They are suitable for wood trim such as doors, windows, and kitchen cabinets, and for large walls and ceilings in kitchens and bathrooms. Semigloss enamels may also be used for priming or undercoating as well as for the finish coat. Two coats should generally be sufficient. Most gloss and semigloss enamels are of the *oil type,* their vehicles consist of drying oils such as varnishes and thinners. Enamel must never be thinned with water.

Flat paints cannot be highly recommended for wood windows, doors, cabinets, and trim, though they may be used on well sealed or well primed wall or ceiling panels of wood or plywood. When used on wood, flat paints should be an oil type and not a water-thinned variety. Any of the flat paints, according to preference, are suitable for plaster, plasterboard, or fiberboard. A wide variety of natural finishes is available for interior woodwork according to the decorative characteristics of the kind of wood and the owner's preference in finish.

On softwoods, staining is usually done with pigmented oil stains. The softer spring wood absorbs much more stain than the summer wood, which may result in a reversal of the normal color gradation in which the summer wood is the darker. To prevent such reversal, the wood may first be sealed with wood sealer and the stain applied as a glaze, which gives a more nearly uniform coloring. On hardwoods, the staining may be done with pigmented oil stains, but clearer coloring is attained with stains made from dyes. Oil stain and so-called nongrain-raising stains require no subsequent sanding of the surface. Water stains raise the wood grain and require light sanding after the stain dries.

The final protective finish may be either a penetrating one, such as linseed oil, one thin coat of shellac, or a wood sealer, or it may consist of two or more coats of varnish. The penetrating finishes may be waxed or not, according to preference. The varnish finishes may be glossy or dull according to the kind of varnish chosen. When desired, a glossy varnish may be rubbed to a medium or low gloss with pumice stone and oil.

PRIMING COAT

The first coat applied to a bare surface is called the *priming coat.* For wood, it consists chiefly of oil and is usually equivalent to a gallon of ordinary paint thinned with a gallon of raw linseed oil. For structural metal the paint is not thinned. In all wood work, nail holes and other defects are filled with putty after the priming coat has been applied.

Exterior Painting

House paints are sold by trade brands, but without any generally recognized commercial standards of quality. Many manufacturers, however, print a statement of the composition on the label. Most manufacturers make one line of house paint that they mark with an identifying trade name, and to which they give a central position in their advertising and in the displays in dealers stores. Since the cost of the paint is nearly always a small proportion of the total cost of painting, the cheaper paints should be avoided and the trade-brand paint should be insisted upon.

There are three main groups of white or tinted house paints that are readily recognized among the available trade-brand paints. They are pure white-lead (L) paint, titanium-lead-zinc (TLZ) paints, and titanium zinc (TZ) paints. Pure white-lead paint contains basic carbonate white lead, linseed oil, drier, and thinner, and nothing else except the necessary tinting colors if the paint is tinted. It is available both in the prepared or ready-mixed form and in the form of soft paste to which its own volume of boiled linseed oil is added to make finish-coat paint; or its own volume of a mixture that is one-half boiled linseed oil and one-half paint thinner to make a priming paint. Differences among the trade-brand white-lead paints are unimportant, except that the prepared paints are oil restricted, whereas the paste paints make oil-rich paint. Although pure white-lead paint usually cost more than TLZ or TZ paints, and usually becomes somewhat grayer from collections of dirt, it holds tints well and is notably trouble free and reliable in performance, particularly when there are uncertainties about the paintability of the wood or the freedom of the structure from so-called moisture conditions.

The trade-brand titanium-lead-zinc (TLZ) paints vary widely in composition. As an indication of the composition considered of good quality, Table 1 shows a TLZ paint that conforms to Federal Specifications.

Paints of lower quality are characterized by decreased content

Table 1. Typical TLZ and TZ Paint Formulas

Ingredients	Content of one gal. of paint		Customary label formula by weight based on	
			Total Paint	Pigment or Liquids
	GALLON	POUNDS	PERCENT	PERCENT
TLZ PAINT				
White lead, basic carbonate	0.030	1.66	11.0	17.6
White lead, basic sulfate	.024	1.28	8.5	13.6
Zinc oxide	.051	2.37	15.7	25.2
Titanium dioxide	.045	1.47	9.8	15.6
Magnesium silicate	.111	2.64	17.5	28.0
Total Pigment	.261	—	62.5	100.0
Raw linseed oil	.437	3.39	22.5	60.0
Boiled linseed oil	.140	1.13	7.5	20.0
Total Nonvolatile	.838	—	—	—
Mineral spirits	.123	.82	5.4	14.5
Liquid paint drier	.039	.31	2.1	5.5
Total Vehicle (Liquid)	—	—	37.5	—
Total Paint	1.000	15.07	100.0	100.0
TZ PAINT				
Zinc oxide	0.63	3.08	22.4	83.0
Titanium dioxide	.024	.77	5.6	9.5
Titanium calcium	.094	2.55	18.6	31.5
Magnesium silicate	.072	1.70	12.4	21.0
Total Pigment	.253	—	59.0	100.0
Raw linseed oil	.363	2.82	20.5	50.0
Boiled linseed oil	.183	1.47	10.7	26.0
Total Nonvolatile	.798	—	—	—
Mineral Spirits	.163	1.07	7.8	19.0
Liquid paint drier	.039	.28	2.0	5.0
Total Vehicle (liquid	—	—	41.0	—
Total Paint	1.0000	13.74	100.0	100.0

of white lead, decreased content of total linseed oil, and increased content of mineral spirits and driers. The TLZ paints are brighter

283

in color, usually stay cleaner, and are cheaper than white-lead paint. Some trade-brand paints and many of the cheaper paints are titanium zinc paints with no white lead at all. Often such paints are called fume resistant because they do not turn black in air contaminated with hydrogen sulfide as do paints containing lead compound. There are a few places in which there is a hazard in the exposure of paints to hydrogen sulfide. The TZ paints are usually cheaper to make than the TLZ paints, but the TZ paints are more exacting about the conditions under which they will give their best performance.

When using pure white-lead paint, the painter may safely tint the white paint on the job to any light color desired. When using TLZ or TZ paints, the paints intended for use as white paint should not be tinted because the colors will soon fade out. Such paints are made with freely chalking kinds of titanium dioxide. For tints, a chalk resistant variety of titanium dioxide is necessary. Accordingly, the painter should purchase TLZ or TZ paint already tinted, or he may buy a so-called white base for tinting supplied for that purpose. The white base for tinting should not be used as white paint because it will not chalk, and it becomes unnecessarily dirty.

The exterior woodwork of a new house should be painted as soon after erection as is practicable. Corrosion-resistant nails, such as dipped galvanized, aluminum, or cadmium plated, are advisable to avoid unsightly rust spots on the paint, particularly if TLZ or TZ paints are to be used. If nails are countersunk, they should not be puttied until after the priming coat of paint has been applied and dried. Wood surfaces should be cleaned and reasonably dry, and any splinters, loose grain, or similar blemishes should be smoothed off, but otherwise no special preparation for painting is necessary.

The first paint job on a new house is the most important one it will ever receive, because the first paint is normally to remain for the useful life of the house as the foundation of all subsequent paintings. New houses are best painted with three coats of paint. Paints should not be thinned excessively by the painter. For two-coat painting, prepared primers and paints should not be thinned at all. For three-coat painting, the finish coat normally should require no thinning, although in cold weather or under some other exceptional circumstances, slight thinning not to exceed one pint of turpentine or mineral spirits to one gallon may be desirable.

One of the major problems in maintenance of a wood covered house is the exterior paint finish. There are a number of reasons for paint failures, many of them known, others not as yet thoroughly investigated. One of the major causes of paint failure is moisture in its various forms. Quality of paint and the method of application are other reasons. Correct methods of application, types of paint, and the problems encountered are covered later in this chapter. Other phases of the exterior maintenance that the owner may encounter in his house are as follows:

1. If steel nails have been used for the application of the siding, disfiguring rust spots may occur at the nailheads. Such rust spots are quite common where nails are driven flush with the heads exposed. Spotting is somewhat less apparent where steel nails have been set and puttied. Similarly, the spotting may be minimized, in the case of flush nailing, by setting the nailheads below the surface and puttying. The puttying should be preceded by a priming coat.

2. Brick and other types of masonry are not always waterproof, and continued rains may result in a damp interior wall or wet spots where water has worked through. If this trouble persists, it may be well to use a waterproof coating on the exposed surface. Transparent coatings can be obtained for this purpose.

3. Caulking is usually required where a change in material occurs, such as that of wood siding abutting against brick chimneys or walls. The wood should always have a prime coating of paint for proper adhesion of the caulking compound. Caulking guns with cartridges can be obtained and are the best means of waterproofing these joints.

4. Rainwater flowing down over wood siding may work through butts and end joints and sometimes may work up under the butt edge by capillary action. Setting the butt end joints in white lead is an oldtime custom which is very effective in preventing water from entering. Painting under the butt edges

at the lap adds mechanical resistance to water ingress. Moisture changes in the siding cause some swelling and shrinking that may break the paint film. Treating the siding with a water repellent before it is applied is an effective method of reducing capillary action. For houses already built, the water repellent could be applied under the butt edges of bevel siding or along the joints of drop siding and at all vertical joints. Excess repellent on the face of painted surfaces should be wiped off.

FINISHING WOOD FLOORS

Wood floors exposed to the weather, such as porch floors, are commonly painted with two coats of porch and deck paint made for the purpose. Generally two coats are needed. Natural finishes are rarely considered durable enough for floors exposed to weather. Concrete floors may be painted, if properly treated.

Interior floors made of hardwoods most often are given a natural finish. There are three kinds of natural finishes widely used on floors. Shellac finish dries rapidly and stays light in color but does not stand exposure to water. Floor-sealer finishes stand wear well because it can be easily renewed at worn places without refinishing the whole floor. Floor varnish provides the most lustrous finish. When shellac is used, three coats are applied and the surface is gone over lightly with steel wool after the first and again after the second coat has been applied. Shellacked floors may be waxed if desired. Wood sealer is usually applied by mopping and is allowed to stand a few minutes before the excess is wiped off. For best appearance and durability it should be sealed and then buffed with steel wool, preferably by using a buffing machine. On newly sanded floors, two coats are generally recommended. The floors may then be waxed if desired. For a varnish finish, three coats of varnish should be applied with sufficient time for each coat to dry before the next one is put on. It is inadvisable to apply shellac for the first coat and varnish for the other because such coatings are likely to show scratches badly. Varnish is characteristically a softer finish and in time dirt and grime may unduly harm the finish resulting in unsightly floor areas.

Oak floor should be filled with wood filler, either natural or colored, before shellac or varnish is applied. When floor-seal finish is used, application of filler is optional. Some forms of wood, laminated wood, or plywood-finish flooring can be purchased with the finish already applied at the factory. Finish of the wood-sealer type is usually chosen for application at the factory. Reasonable care should be taken after such flooring has been laid in a new house to protect it from heavy traffic incident to completion of the house.

PLYWOOD AND WALLBOARD

The painting and finishing characteristics of plywood are essentially those of the kind of wood with which the plywood is faced. In large sheets of plywood, any wood checking is more objectionable in appearance than it is on boards of lumber. Plywood with hardwood faces usually is not much more prone to checking than lumber of the same species, but plywood faced with most softwoods is likely to check to some extent even when used indoors with protective finishes. Such checking is more conspicuous with some finishes of light color, but somewhat less as the gloss diminishes and also as the color is darkened greatly.

With opaque finishes, the checking is least objectionable when flat finishes with rough surfaces are chosen, such as stippled, textured, or sanded finishes. Checking is less readily observable with natural finishes than with opaque finishes, because the grain pattern of the wood distracts attention from the checking. Softwood plywood may be used without danger that checking will mar the finish if the exposed face of the plywood is covered with paper, cloth, or resin-impregnated paper firmly glued in place. For exterior exposure, of course, the glue must be thoroughly weather resistant. Plasterboard and fiberboard are, for the most part, paintable with the same material and methods that are suitable for plaster.

The use of sheet materials for covering walls and ceilings poses problems at the joints between the sheets. Sometimes the joints are deliberately left evident as part of the decorative scheme. Or, they may require special treatment, such as taping to conceal the joints and to permit a smooth covering of paint. Manufacturers of the sheet materials usually furnish detailed directions or suggestions for accomplishing the desired results.

CHARACTERISTICS OF WOODS FOR PAINTING

Some woods allow a wide freedom in choice of paints and painting procedures, whereas others are more exacting in their requirements if fully satisfactory experience is to be obtained. Woods differ also in the appearance of natural finish obtainable with them in the ways in which they behave if exposed to the weather without protective coating or treatment. Woods that offer the greatest freedom of choice for painting are the woods that are low in density, slowly grown, cut to expose edge grain rather than flat grain on the principal surface, are either free from pores or have pores smaller than those in birch, and are free from defects such as knots and pitch pockets. Flat-grain boards hold paint better on the back side than the pitch side. All of the pines, spruces, and Douglas-firs are improved for painting by kiln-drying to fix the resin in the wood so it will not exude through the coating.

Eastern red cedar and Spanish cedar contain slowly volatile oils that prevent proper drying of most paints and varnishes unless most of the oils are removed by exceptionally thorough kiln-drying. Incense cedar, Port-Oxford cedar, and cypress may sometimes give similar troubles if they contain too much of their characteristic oils. The method of seasoning on most other woods has little or no effect on paintability. Water-soluble extractives in redwood, western red cedar, walnut, chestnut, and oak may retard drying of coatings applied while the wood is damp or wet and may discolor paint if the wood becomes thoroughly wet after painting. Otherwise, such extractives contribute to better durability of paint.

On exterior surfaces, softwoods that expose wide bands of summer wood require extra care in painting because it is over such bands of summer wood that the paint coating, when embrittled by the action of the sun and rain, begins to wear away by crumbling or flaking. Repainting is usually considered before too much summer wood becomes exposed. Coatings on interior surfaces seldom wear away by crumbling or flaking. But greater smoothness of painted surfaces may be demanded on interior surfaces. The presence of wide bands of summer wood makes it necessary to take extra care in smoothing the surface before painting and in applying undercoats to keep the wood from showing ridges in the coating, commonly called raised grain.

Sometimes the easiest woods to paint smoothly may be considered too soft and too easily dented to be a good choice for interior woodwork, in which case a compromise must be reached between the properties that favor smooth painting and those that give enough hardness for the intended use. For good service of paints or other finishes, wood must be kept reasonably dry, in which case it is not subject to decay. But there are some places, such as outside steps, railings, fence post, porch columns, and exterior paneled doors, where joints in the structure may admit enough water to permit stain or decay to develop in nondurable kinds of wood. In such cases the performance of both wood and paint may often be improved by treating the wood with preservatives or water-repellent preservatives before painting. Treatment of the wood with water-repellent preservatives at least several days before erection has been found effective in minimizing such penetration of water.

PAINTING TOOLS

A satisfactory paint job requires cleaning, sanding, and scraping. To prepare a surface for painting, the proper tools should be on hand and in good condition and ready for use. These tools range from putty knifes to various sizes of paint brushes.

Knives

There are various types of knives which include, palette, stone, and stopping. The palette knife is a long and very flexible knife with a blunt edge and rounded at the end. This knife is used for mixing colors, or scraping paint off of the palette. In using the palette knife, it should be held as flat as possible to avoid making indentations in the palette. The stone knife is similar in shape to the palette knife but it is larger and is used in connection with the slab in mixing colors. The stopping knife is used for stopping cracks, etc., with the use of putty. This knife is shorter in proportion than the palette knife and is spear shaped, broad, and stiff.

289

It is adapted to bear the force necessary to press the putty into the crevices.

Brushes

Paint brushes used to apply paint are made in various sizes and shapes. The larger brushes are called *pound brushes,* which are termed *four 0, six 0, and eight 0.* These brushes are made both round and flat. Paint brushes are bound either with string or with copper wire. They are sometimes used as dusters before being used in paint, but this is not recommended. The smaller brushes are called *tools* or *sash tools,* and may be obtained in about a dozen different sizes; some are bound with string and others fixed in metal. The smallest hog-hair brushes are called *fitches,* and are used for small work where tools would be too large. The smallest brushes are the camels hair, with long or short hair, according to the work to be done. Varnish brushes are made in different sizes and are also made flat and with different quality of hair. A dusting brush for removing dust from the surface to be painted has

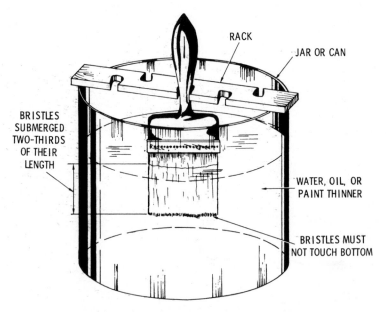

RACK

JAR OR CAN

BRISTLES SUBMERGED TWO-THIRDS OF THEIR LENGTH

WATER, OIL, OR PAINT THINNER

BRISTLES MUST NOT TOUCH BOTTOM

Fig. 3. A method of suspending a brush in a can of water or oil.

longer hairs than a regular paint brush. The hairs are bound so they will spread out when used.

CARE OF PAINT BRUSHES

When not in use the paint brush should be wiped off and suspended in a pail or can with the bristles two-thirds submerged in water or oil, as shown in Fig. 3. If the brush bristles rest on the bottom of the pail, the bristles will curl and damage the brush (Fig. 4). Brushes cannot be properly cleaned in gasoline, especially if it has been used in colors containing turpentine. The turpentine will form a mass that will cling to the bristles. One method is to clean the brush in benzine after use and wrap it up in a linseed oil soaked paper for storage. This will keep the bristles limber for use later. Another bad feature with bristles is the swelling or enlargement caused by allowing the brush to remain suspended in water or oil too long, as shown in Fig. 5. When a brush is not to be used for a long interval of time it should be thoroughly cleaned and put away.

TREATMENT OF HARDENED BRUSHES

Soak the hardened brush for 24 hours in linseed oil or until the bristles soften. Then place the brush in a can of benzine until the bristles are soft enough to work the paint loose. There are paint-

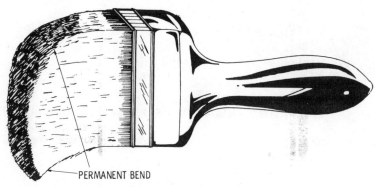

PERMANENT BEND

Fig. 4. Permanent distorted bristles, caused by brush resting on bottom of can.

291

brush cleaners on the market which will clean brushes that have had hardened paint on them for years.

Another method for cleaning paint brushes is shown in Fig. 6. The brushes are suspended above the lacquer thinner in a closed can. The fumes from the thinner will liquefy the paint in the bristles and drip down into the thinner. This point is proven by the removal of the paint on the brush handle.

Fig. 5. Swelling or enlargement of bristles.

ESTIMATING QUANTITY OF PAINT REQUIRED

An approximate calculation of the quantity of paint required to paint a house may be made as follows

Measure the girth of the house, and the height to the eaves. If there are gables, measure half-way up and multiply the

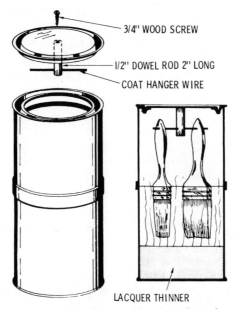

3/4" WOOD SCREW

I/2" DOWEL ROD 2" LONG

COAT HANGER WIRE

LACQUER THINNER

Fig. 6. Suspending the paint brushes above lacquer thinner in a closed can.

height by the length which will give the number of square feet. Divide the square feet by 600 which will determine the approximate number of gallons. If a second coat is required, double the quantity. If you mix the paint, the average formula per gallon is:

Pure white lead.................................... 15-1/2 pounds
Pure linseed oil.................................... 4-1/2 pints
Pure turpentine.................................... 1/2 pint
Pure drier... 1/2 pint or less

The painter must use his judgement, or make tests, as to the conditions of the surface to be painted, the season, etc., which may call for more or fewer ingredients.

SPRAY PAINTING

With the construction of modern equipment, spraying has be-

293

come an increasingly popular method of applying paint to all types of interior and exterior surfaces. It is much faster than brushing, both in application and in the use of faster drying finishes. Small spraying units of the better types are very efficient, performing perfect work with paint, lacquer, varnish, and synthetics.

Spray equipment used around the home and in the average workshop consist of three items which are:

1. Air compressor.
2. Gun.
3. Hose.

Air Compressor

There are two kinds of air compressors—the *piston* type and the *diaphragm* type. The piston type consists essentially of a metal piston working inside a cylinder, very much like the piston in an automobile engine. On the down-stroke of the piston, air is drawn into the cylinder and on the up-stroke, this air is compressed. A

Fig. 7. A typical 1/3 horsepower stationary paint sprayer.

piston-type stationary paint sprayer is shown in Fig. 7, and a portable type shown in Fig. 8.

The diaphragm-type compressor differs from the piston type mainly in that it is equipped with a rubber diaphragm taking the place of the piston, as shown in Fig. 9. Because of the fact that the diaphragm type of compressor is rather inexpensive and also requires less maintenance, it is usually preferred in the spraying applications that are performed around the home where the air requirements are small.

The motor supplying the power for the compressor is furnished in fractional horsepower, and it also determines the size of the compressor. For home spraying, 1/4 to 1/3 horsepower is the smallest motor practical to use for satisfactory spraying.

Air Volume and Pressure

The basic requirements for any paint sprayer is an efficient high-output compressor that supplies the spray gun with a smooth

Fig. 8. A portable 1/2 horsepower paint sprayer.

295

continuous flow of air, permitting the paint to be applied without excessive thinning, and providing a smooth even coat. The measure of performance of any paint sprayer is the smooth even application of paint to any surface being painted.

The following terms are used to measure the performance of a compressor:

Air Displacement—This is the theoretical amount of air in cubic feet the compressor can pump in one minute at working pressure. It is a relatively simple matter to calculate the air displacement of a compressor if the cylinder diameter, length of stroke, and rpm is known. For example, the area of the cylinder multiplied by the length of the stroke and the shaft revolutions per minute, equals the displacement volume. The formula for computing it is as follows:

$$\frac{\text{Area of cylinder} \times \text{stroke} \times \text{rpm} \times \text{number of cylinders}}{1,728}$$
$$= \text{Displacement in CFM}$$

Both cylinder diameter and stroke measurement in the foregoing formula should be in inches. The formula applies to single-stage compressors, but can also be used for the two-stage compressor as well.

Air Delivery—An air compressor is rated on the actual amount of free air pumped in one minute at working pressure for use by the spray gun. Air delivery is always less than the displacement rating, because no compressor is 100% efficient. *Air delivery at working pressure, and not the displacement or the horsepower, is the true rating of the compressor.* A piston or diaphragm compressor powered with a 1/3-horsepower motor will deliver about 3 cubic feet of free air per minute. Many dealers in paint spraying equipment never mention air delivery, but feature only terms of air displacement, which is the larger figure.

Working Pressure—Working pressure is the number of pounds of pressure a compressor will maintain while in continuous operation with a specific spray gun. Each gun requires a constant air delivery at a certain working pressure for maximum efficiency.

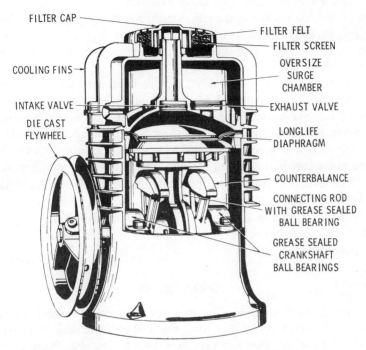

FILTER CAP

FILTER FELT

FILTER SCREEN

OVERSIZE SURGE CHAMBER

COOLING FINS

INTAKE VALVE

EXHAUST VALVE

DIE CAST FLYWHEEL

LONGLIFE DIAPHRAGM

COUNTERBALANCE

CONNECTING ROD WITH GREASE SEALED BALL BEARING

GREASE SEALED CRANKSHAFT BALL BEARINGS

Fig. 9. A typical interior view of a diaphragm type air compressor.

Too much air pressure causes over atomization and results in a splatter-like finish.

Pressure and Exhaust Opening Relations—The relationship between the exhaust opening of a spray gun and pressure is such that when the size of the exhaust opening is increased, the pressure will decrease and when the opening is reduced, the pressure will increase. This inverse pressure and exhaust opening relationship may readily be observed if it is noted that an average 1/3-horsepower compressor will deliver about 40 pounds of free air through an exhaust opening of about 1/16 inch in diameter. If the opening is closed, the pressure will immediately jump to about 150 pounds or more, resulting in motor stoppage due to operation of the automatic pressure switch, or other safety device incorporated in the equipment.

Air Pressure and Air Hose relations—When pumping air from the compressor to the gun, a pressure drop is produced. This pressure drop is due to friction between the flowing air and the walls of the hose, pipe line, and passage through which the air must travel.

Too often a spray gun is blamed for improper functioning, or a material is considered of inferior quality, when the real trouble is an inadequate supply of compressed air at the gun. Frequently, operators believe they are using very high pressure when an investigation will reveal that, due to a small hose size or one of excessive length, the pressure is inadequate for proper atomization.

The pressure drop is an important factor to consider and is shown in Table 2. The pressure drop shown in the table is based on the air consumption of a spray gun equipped with an air gap drawing approximately 12 cubic feet of air per minute at 60 pounds pressure.

SPRAY GUNS

In operation, the spray gun atomizes the material which is being sprayed and the operator controls the flow of the material with the trigger attachment on the gun, as shown in Fig. 10. Spray guns are supplied with various sizes of nozzles and air gaps to permit the handling of different types of material, as shown in Fig. 11. An air adjustment built into the gun makes possible a change in the shape and size of the spray pattern, ranging from a wide fan-shaped spray to a small round spray as the occasion requires.

Spray guns are either *external mix* or *internal mix*. In the external-mix type of gun, the material leaves the gun through a hole in the center of the cap, while air from an annular ring surrounding the fluid stream (as well as from two or more holes or orifices set directly opposite each other) engages the material at an angle. Spray guns in which the air and material are mixed before leaving the orifice of the gun are of the internal-mix type. The suction-feed gun, as the name implies, syphons the material from a cup attached to the gun and is generally used for radiators, grills, and other small projects which require light bodies and only a small amount of material.

Table 2. Air Pressure at Spray Gun

Size of Air Hose (inside diameter)	Air Pressure Drop at Spray Gun					
	5-foot length	10-foot length	15-foot length	20-foot length	25-foot length	50-foot length
1/4-inch	Lbs.	Lbs	Lbs.	Lbs.	Lbs.	Lbs.
At 40 lbs. pressure	6	8	9½	11	12¾	24
At 50 lbs. pressure	7½	10	12	14	16	28
At 60 lbs. pressure	9	12½	14½	16¾	19	31
At 70 lbs. pressure	10¾	14½	17	19½	22½	34
At 80 lbs. pressure	12¼	16½	19½	22½	25½	37
At 90 lbs. pressure	14	18¾	22	25¼	29	39½
5/16-inch						
At 40 lbs. pressure	2¼	2¾	3¼	3½	4	8½
At 50 lbs. pressure	3	3½	4	4½	5	10
At 60 lbs. pressure	3¾	4½	5	5½	6	11½
At 70 lbs. pressure	4½	5¼	6	6¾	7¼	13
At 80 lbs. pressure	5½	6¼	7	8	8¾	14½
At 90 lbs. pressure	6½	7½	8½	9½	10½	16

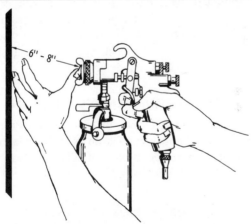

Fig. 10. A typical paint spray gun.

The pressure-feed gun forces the material by air pressure from the container. Air directed into the container puts the material under pressure and forces it to the nozzle. The pressure-type gun is undoubtedly the best all-around gun for use with small compressors and with such materials as house paint, floor enamels, wall paints, and the like. Any spray gun is either a bleeder or non-bleeder type. A bleeder type is one which is designed for passage

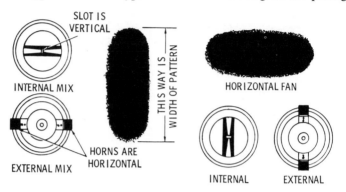

Fig. 11. Various types of nozzles used to permit different patterns of spray.

of air at all times, whereas a nonbleeder gun cannot pass air until the trigger is pulled. This nonbleeder type of gun is used only when air is supplied from a tank or from a compressor equipped with pressure control. Bleeder guns are always used when working directly from small compressors, such as those used in small workshops or for spraying jobs around the home.

SPRAYING EQUIPMENT CONNECTIONS

There are several possible hook-ups for various types of applications. The simplest possible hook-up is one where the feed is directly from the compressor to the gun. This is a preferred spray method on large surface paint jobs.

AIR COMPRESSOR ACCESSORIES

One of the most useful accessory for home use spraying is the angle head. An angle head fitted to the pressure gun prevents excessive tilting of the fluid cup when spraying floors, ceilings, and similar surfaces. A respirator is seldom required for work of short duration, but it is a very useful item when spraying for long periods in confined quarters.

Another useful accessory is the condenser which filters water, oils, and various foreign particles out of the air supply. If the condenser is fitted with an air regulator, the combination is termed a transformer. With a condenser-regulator combination, it is possible to have the air filtered, while the regulator allows setting the air pressure at a predetermined rate as required for the particular job. Additional useful accessories consist of an air-duster gun and extra lengths of air hose with couplings to facilitate an increase in the working radius.

SPRAY BOOTHS

Where spray equipment is installed in a permanent location and the work to be sprayed is of such a size that it can be conveniently done in one place, exhaust equipped spray booths are usually provided for the removal of vaporized paint and odors. Spray booths are made in various sizes and types for specific purposes, ranging from a simple frame and plywood or sheet-metal enclosure to well equipped commercial-type water-wash booths used in production plants. Fig. 12 illustrates a typical corner spray booth.

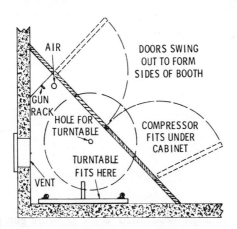

Fig. 12. A typical corner spray booth suitable for the average home workshop.

The ordinary floor-type booth may accumulate a certain amount of waste material after the spraying has been completed. These should be removed immediately, since cleanliness is important to good spray results. When spraying is confined to an area, it is advisable to have a portable fan secured in a frame which can be placed in a window to provide proper ventilation.

Turntables of simple construction are used in spray booths to facilitate rotation of the work while spraying. A suitable turntable eliminates a lot of body movement as well as handling of the work during the spraying process.

PLACING THE WORK

Prior to the actual spraying, it is necessary to arrange the work properly, and one of the fundamental rules relating to placement is to get the work off the floor. There are several methods to accomplish this, the most common of which is to provide a couple of sawhorses bridged with suitable boards or planking. Another method to get the work off the floor and in a comfortable spraying position is to suspend the work with a rope or hooks from a clothesline or other convenient overhead support. This suspension method offers maximum convenience in manipulating objects that are without a base.

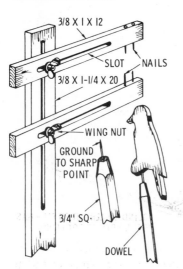

Fig. 13. Holding device for spraying small articles of wood or other light material.

Small wooden articles can be readily sprayed if speared on an icepick, or other special holding device, such as illustrated in Fig. 13.

For general indoor spraying, an adjustable backstop stand, such as shown in Fig. 14, will prove very satisfactory. This stand uses newspapers or paper rolls and provides a suitable surface for testing the spray gun as well as a backstop when spraying the work. Pedestal turntables with adjustable heights are good for placement of small furniture and similar articles to be sprayed. Various base styles may be used to suit individual preferences, as shown in Fig. 15. Other practical supports and work-holding devices may be improvised to suit almost any condition and space requirement. Large articles of wood or metal are sometimes suspended by a block-and-tackle arrangement, permitting the work to be hoisted out of the way for drying after spraying.

SPRAYING TECHNIQUE

Prior to commencing the actual spraying job, it is good practice to test spray a few panels in order to obtain the feel of the gun

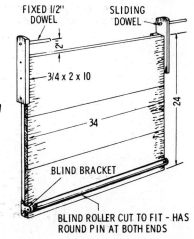

FIXED 1/2" DOWEL SLIDING DOWEL

3/4 x 2 x 10

24

34

BLIND BRACKET

BLIND ROLLER CUT TO FIT - HAS ROUND PIN AT BOTH ENDS

Fig. 14. Typical backstop stand suitable for spray pattern.

and to make any necessary adjustments. Such practice spraying can be done on old cartons or on sheets of newspaper tacked to a carton or box.

303

With the spray cup filled with water-mixed paint or other inexpensive material, practice spraying can be done directly on any odd material, as previously noted. It will be useful to experiment with the full range of fluid adjustment, starting with the fluid needle screw backed off from the closed position just far enough to obtain a small pattern an inch or so wide when the trigger is pulled all the way back. The spray-gun stroke is made by moving the gun parallel to the work and at right angles to the surface. The speed of stroking should be about the same as brushing. The distance from gun to work should be between 6 and 8 inches, as

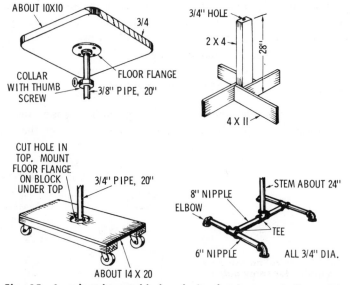

Fig. 15. A pedestal turntable for placing furniture or similar articles for spraying.

shown in Fig. 16. Practice should be done with straight uniform strokes moving back and forth across the surface in such a way that the pattern laps about 50 percent on each pass. An appreciable variation of distance between gun and work or angle of the gun will result in uneven coatings. Therefore, always hold the gun without tilting and move it correctly. Spraying is merely the action of the wrist and forearm and the proper technique is readily and easily acquired. Fairly large pieces of scrap wood remaining from the project are excellent for practicing spray patterns, because this gives the operator an idea of any problems he might expect in actually spraying the final project itself.

Release the trigger at the end of each stroke to avoid piling up material. Avoid pivoting or circular movements of the wrist or forearm or any oblique spraying which will cause material to rebound from the surface. It causes excessive mist and waste material as well as a patchy, unsatisfactory job.

Cleaning the Gun—To obtain continued and satisfactory service, a spray gun (like any other tool) requires constant care. It should not be left in a thinner overnight, because such practice removes the lubricant from the packing through which the needle moves, thereby causing the needle to stick and the gun to "spit." A dirty air gap causes a defective spray. Thus, it is good practice to clean it promptly after the spraying job is finished. At the end of the day, remove the gun from the cup and hose, and thoroughly clean it. Clean the entire gun, the outside as well as the inside. The cup or tank should likewise be cleaned.

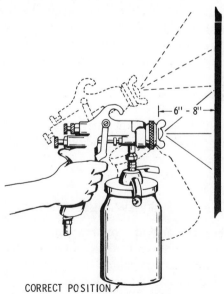

CORRECT POSITION

Fig. 16. Illustrating the correct position of spray gun when spraying.

305

SUMMARY

The base materials used in producing paints are pigments, white lead, sublimed white lead, zinc white, barytes, whiting, vehicle, and linseed oil. Paint is a mixture consisting of finely ground substances, called pigments, held in suspension in a liquid called a vehicle.

The three primary colors are red, blue, and yellow. The combinations of these three primary colors produce what are known as secondary colors, which are purple, green, and orange. By mixing any two of the secondary colors together you get what are known as tertiary colors, which are olive, citrine, and russet. Black and white are not generally regarded as colors. A good black can be produced by mixing all three primary colors together in the proper proportions.

Paint brushes should always be throughly cleaned after each usage. When not in use, the brush should always be wiped off and suspended in a pail or can with the bristles two-thirds submerged in water or oil. Never allow the bristles to rest on the bottom of the pail; the bristles will curl and damage the brush. Hardened paint can be removed from brushes in various ways. There are paintbrush cleaners sold on the market which will clean brushes which have had paint embedded in the bristles for years. If a brush is not to be used for a long period of time, it is best to thoroughly clean the brush and put it away.

Spray painting has become an increasingly popular method of applying paint to all types of interior and exterior surfaces. It is faster than brush painting, and generally produces a much smoother finish.

REVIEW QUESTIONS

1. Name the material used to produce paint.
2. Name the three primary and three secondary colors.
3. Name the three tertiary colors.
4. How should paint brushes be cleaned?
5. What are a pound brush, sash tool, and fitches?

CHAPTER 21

Maintenance and Repair

A house that is well constructed, with adequate attention to construction details and to the choice of materials as brought out in the preceding chapters in the book, will need far less maintenance than one that is not well built. The timeworn phrase, "It's not the first cost, it's the upkeep," certainly applies to a house that requires more maintenance than its well-constructed neighbor. It is indeed discouraging to the houseowner to begin repair and maintenance almost before he has moved in! An extra $10 used for rust-resistant nails on the siding, for example, may save $100 by requiring less frequent painting.

Inasmuch as there are good publications available on the methods of repairing and caring for the house, this section will deal principally with ways of foreseeing possible future trouble spots. A small amount of initial attention will often prevent a major repair bill.

BASEMENT

The basement may sometimes be damp for several months after the house has been completed. In most cases, this moisture comes from masonry walls and floors and will progressively disappear. In cases of persistent dampness, however, the owner should check various areas in order to eliminate any possibilities for water entry.

The following areas may be the source for some of the trouble caused by the entry of moisture:

1. Check the drainage at the downspouts. The final grade around the house should be away from the building, and a splash block should be provided to drain water away from the foundation wall.
2. Some settling of the soil may occur at the foundation wall and form water pockets. These areas should be filled and tamped so that surface water can drain away.
3. Some leaking may occur in a poured concrete wall at the tie wires. These usually seal themselves, but larger holes should be filled with a cement mortar. Clean and slightly dampen the area first for good adhesion.
4. Concrete-block or other masonry walls exposed above grade often show dampness on the interior after a prolonged rainy spell. There are concrete paints on the market that increase the resistance to moisture seepage. Transparent liquid waterproofing materials may also be used on the exterior walls.
5. There should be at least a 6-inch clearance between the bottom of the siding and the finish grade to prevent moisture absorption. Shrubs and foundation plantings should also be kept away from the wall to improve circulation and drying. In lawn sprinkling, do not allow the water to spray against the walls of the house.
6. Check areas between the foundation wall and the sill. Any openings should be filled with a cement mixture or a caulking compound. This filling will decrease heat loss and also prevent entry of insects into the basement.
7. Dampness in the basement in the early summer months is often augmented by opening the windows for ventilation during the day. This will allow warm, moisture-laden outside air to enter. The lower temperature of the basement will cool the incoming air and frequently cause condensation to collect and drip from cold-water pipes and also collect on colder parts of the masonry walls and floors. To air out the basement, open the windows during the night.

In aggravated cases, heating the basement to raise the temperature 5 to 10 degrees, with the windows closed, has proved helpful. Heating lowers the humidity, warms the cold masonry, and dries up the moisture. On cool days no heat is required, and the windows may be opened. As the summer advances and the masonry warms up, condensation will not usually occur on the walls but may occur on the cold-water pipes. Wrapping such pipes has proved helpful in reducing condensation and drip.

A desiccant, such as calcium chloride, is sometimes used to lower the humidity in basements. Mechanical dehumidifiers are also available in a variety of sizes. Such units are generally enclosed in a cabinet and are operated by electricity. The doors and windows should be closed at all times during dehumidification; otherwise the moisture in the entering air will continually replace the moisture extracted by the desiccant or dehumidifier.

CRAWL-SPACE AREA

The crawl-space area should be checked as follows:

1. If the house is located in a termite area, be sure to make annual inspections, preferably in the late spring or early summer, for termite activity or damage. These inspections are a must if the house or a part of the house is built over a crawl space. Termite tubes on the walls or piers are an indication of their activity. A well-constructed house will have a termite shield under the wood sill with a 2-inch extension on the interior. Examine the shield for proper projection, and also any cracks in the foundation walls, as these cracks form good channels for termites to enter.

2. While in this crawl space, it is well to check the area for any decay that may be occurring in the girders or joists as well as for signs of condensation on wood members of the floor framing. Use a penknife to test questionable areas for rot and decay.

3. If the crawl space seems damp and the ground moist, it may be due to lack of proper ventilation. The crawl-space ventilators should be of adequate size and should be so located that there is cross circulation. The use of a soil cover, 55-pound saturated felt or heavier, on the ground of the crawl space will prevent much of the soil moisture from entering the area.

ROOF AND ATTIC

The roof and attic should be inspected in the following respects:

1. If there are a few humps on asphalt-shingle tabs, they most likely are due to nails that have not been driven into solid wood. Remove such nails and replace with others driven into sound wood. There may be a line of buckled shingles along the roof. This ordinarily is caused by applying the shingles to roof boards that are not dry and to roof sheathing that varies in thickness. The shrinkage of the wood in the wide boards between the nails will also cause these humps by buckling the shingles. Time and hot weather tend to reduce this condition.

2. A dirt streak down the gable end of the house is often caused by rain entering the rake moulding from shingles that have insufficient projection. A cant strip would ordinarily have prevented this situation.

3. In winters of heavy snows, ice dams may form at the eaves, and these often result in water entering the cornice and walls of the house. The immediate remedy is to remove the snow on the roof for a short distance above the gutters and, if necessary, in the valleys. Additional insulation between heated rooms and roof space, and increased ventilation will help to decrease the melting of snow on the roof and thus minimize formation of ice at the gutters and in the valleys. Deep snow in valleys also sometimes forms ice dams that cause water to back up under shingles and valley flashing.

4. Roof leaks are often caused by improper flashing at the valley, ridge, or around the chimney. Observe these areas during a rainy spell to discover the source. Water may travel many feet from the point of entry before it drips off the roof members.

5. The attic ventilators serve two important purposes—that of summer ventilation as a means of lowering the attic temperature to improve comfort conditions in the rooms below, and that of winter ventilation to remove moisture that may work through the ceiling and condense in the attic space. The ventilators should be open both winter and summer.

In order to check for sufficient ventilating area during cold weather, examine the attic after a prolonged cold period. If nails protruding from the roof are heavily coated with frost, it is evident that ventilation is not sufficient. Frost may also collect on the roof sheathing, first appearing near the eaves on the north side of the roof. Increase the size of the ventilators or place additional ones in the protected underside of the cornice. This will improve air movement and circulation.

SELECTION OF LUMBER

The grading of lumber is a complex matter, and only during the last few years has it been simplified so as to provide the possibility of uniformity over the entire country. In this chapter it is intended to provide the repair man with the fundamentals of lumber grading so that he can know the subdivisions of grading rules and where to secure further information on the subject if desired.

For many years the lumber industry has been faced with a variety of trade practices in nomenclature, grades, and sizes of lumber. Individuals felt satisfied that lumber, having similar characteristics and intended to be used for similar purposes, even though cut from different species, could be manufactured, sold, and used according to definite standards. Conferences have been held with the results that representatives of manufacturers, distributors, wholesalers, retailers, users of lumber, architects, engineers, and general contractors, through the Bureau of Standards of the U. S. Department of Commerce, recommended classification of nomenclature, basic grades, specific sizes, certain descriptions, measurements, tallies, shipping provisions, grade marking, and inspection of softwood lumber be established as standards.

The two major objectives were attained through these series of meetings as follows:

1. The actual finished yard-lumber items were reduced nearly 60% in number, thus eliminating a large percentage of unnecessary and wasteful sizes.
2. The builders of homes in the United States can now be assured that the industry as a united force will give them standard lumber and standard products.

311

Because it is based upon the judgment of the grader and a visual inspection of each piece, lumber grading cannot be expected to reach scientific accuracy. However, it is estimated that by the rules set down a variation limited to 5% can be expected in the results of careful grading by different individuals. With the exception of dimension material, yard-lumber grade is determined (when rough or surfaced two sides) from the better or face side of the piece. Lumber, however, when surfaced one side only is graded from that side.

Defects or blemishes, or a combination of defects and blemishes, which are not described in the standard grading rules are considered according to a comparison with known defects and are graded on a basis of the damaging effect upon the pieces of lumber being graded. When the angle is 45° or more between the surface of the piece and the rings (so-called grain), the material is considered *edge grain*. If less than 45° at any point the material is known as *flat (slash) grain*.

CLASSIFICATION FOR USE OF LUMBER

American lumber standards specify that lumber be classified according to its principal uses:

1. Yard lumber.
2. Structural timber.
3. Factory or shop lumber.

Light frame construction concerns itself to a very minor extent with other than yard lumber; therefore no discussion is included for this type of lumber, except to state that structural timbers are 5 inches or over in thickness and width, and are graded according to the strength and the use of the entire piece. Factory or shop lumber, however, is lumber intended for further manufacture and is graded on a basis of the proportion of area for the production of a limited number of cuttings of specified given minimum size and quality.

Yard lumber is generally less than 5 inches in thickness and is intended to be used for general building purposes. Unlike factory or shop lumber, yard lumber is graded upon the use of the entire piece.

Yard lumber has classification by dimensions as follows:

1. Strips which are less than 2 inches thick and not over 8 inches wide.
2. Boards which are less than 2 inches thick and 8 inches or more in width.
3. Dimensions which consist of all yard lumber with the exception of boards, strips, and timber.

Yard lumber which is between 2 and 5 inches thick and of any width comes under the classification of dimension:

1. Planks are yard lumber which is at least 2 inches but less than 4 inches in thickness, and 8 inches or over in width.
2. Scantlings are yard lumber which is at least 2 inches but under 5 inches in thickness, and less than 8 inches wide.
3. Heavy joists are yard lumber which is at least 4 inches but less than 6 inches in thickness, and 8 inches or over in width.

In respect to yard-lumber quality, there are two main divisions:

1. **Select Lumber**—Select lumber, according to the American Lumber Standards, is:

    Generally clear, containing defects limited both as to size and number, and smoothly finished and suitable for use as a whole for finishing purposes or where large clear pieces are required.

    Two classes of select lumber are provided:
    (a) That suitable for natural finishes, graded as A and B.
    (b) That having defects and blemishes of somewhat greater extent than (a), but which can always be covered by paint, graded as C and D.

2. **Common Lumber**—The American Lumber Standards specify common lumber as that having numerous defects and blemishes preventing its use for finishing purposes, but suitable for general utility and construction purposes.

    Again, American Lumber Standards recognize two general classes:

313

    (a) Common lumber which can be used as a whole for purposes in which surface covering or strength is required. All defects and blemishes in this classification must be sound. This class includes No. 1 and No. 2 common.

    (b) Common lumber containing very coarse defects likely to cause waste in the use of the piece. This class includes No. 3, No. 4, and No. 5 common.

Dimension lumber is available only in the three upper grades, No. 1, No. 2, and No. 3 common.

The outline in Table 1 is a key to the American Lumber Standards.

EXTERIOR WALLS

One of the major problems in maintenance of a wood-covered house is the exterior paint finish. There are a number of reasons for paint failures, many of them known, others not as yet thoroughly investigated. One of the major causes of paint failure is moisture in its various forms. Quality of paint and method of application are other reasons. Correct methods of application, types of paint, and the problems encountered are covered under Chapter 18, on Painting and Equipment.

Other phases of the exterior maintenance that the owner may encounter in his house are that if steel nails have been used for the application of the siding, disfiguring rust spots may occur at the nailhead. Such rust spots are quite common where nails are driven flush with the heads exposed. Spotting is somewhat less apparent where steel nails have been set and puttied. The spotting may be minimized, in the case of flush nailing, by setting the nailhead below the surface and puttying. The puttying should be preceded by a priming coat. Take care of these nails just before the house requires repainting.

Brick and other types of masonry are not always waterproof, and continued rains may result in damp interior walls or wet spots where water has soaked through. If this trouble persists, it may be well to use a waterproof coating over the exposed surfaces. Transparent coatings can be obtained for this purpose. Caulking is usually required where a change in materials occurs, such as that of wood siding abutting against brick chimneys or walls. The wood should always have a prime coating of paint for proper adhesion of the caulking compound. Caulking guns with cartridges can be obtained and are the best means of waterproofing these joints.

Rainwater flowing down over wood siding may work through butt and end joints and sometimes may work up under the butt edge by capillary action. Setting the butt end joints in white lead

Table 1. Base Grade Classification for Yard Lumber

TOTAL PRODUCTS OF A TYPICAL LOG ARRANGED IN SERIES ACCORDING TO QUALITY AS DETERMINED BY APPEARANCE

COMMON LUMBER CONTAINING DEFECTS OR BLEMISHES WHICH DETRACT FROM A FINISH APPEARANCE, BUT WHICH IS SUITABLE FOR GENERAL CONSTRUCTION

LUMBER PERMITTING WASTE

NO. 3 COMMON-ALLOWS LARGER AND COARSER DEFECTS THAN NO. 2 AND OCCASIONAL KNOT HOLES

NO. 4 COMMON-LOW QUALITY LUMBER ADMITTING THE COARSEST DEFECTS SUCH AS DECAY AND HOLES

NO. 5 COMMON - MUST HOLD TOGETHER UNDER ORDINARY HANDLING

LUMBER SUITABLE FOR USE WITHOUT WASTE

NO. 2 COMMON-ALLOWS LARGE AND COARSE DEFECTS. MAY BE CONSIDERED GRAIN-TIGHT LUMBER

NO. 1 COMMON-SOUND AND TIGHT KNOTTED. SIZE OF DEFECT AND BLEMISH LIMITED. MAY BE CONSIDERED WATERTIGHT

SELECT LUMBER OF GOOD APPEARANCE AND FINISHING QUALITIES

SUITABLE FOR A PAINT FINISH

GRADE D - ALLOWS ANY NUMBER OF DEFECTS, BUT DOES NOT DETRACT FROM A FINISH APPEARANCE, ESPECIALLY WHEN COVERED WITH PAINT

GRADE C - ALLOWS LIMITED NUMBER OF SMALL DEFECTS OR BLEMISHES WHICH CAN BE COVERED WITH PAINT

SUITABLE FOR NATURAL FINISH

GRADE B - ALLOWS A FEW DEFECTS OR BLEMISHES

GRADE A - PRACTICALLY FREE FROM DEFECT

is an oldtime custom that is very effective in preventing water from entering. Painting under the butt edges at the lap adds mechanical resistance to water ingress. However, moisture changes in the siding cause some swelling and shrinking that may break the paint film. Treating the siding with a water repellent before it is applied is an effective method of reducing capillary action. For houses already built, the water repellent could be applied under the butt edges of bevel siding or along the joints of drop siding and at all vertical joints. Excess repellent on the face of painted surfaces should be wiped off.

INTERIOR WALLS

In a newly constructed house, many small plaster cracks may develop during or after the first heating season. These cracks are usually due to the drying and shrinking of the structural members. For this reason, it is advisable to wait for a part of the heating season before painting plaster. These cracks can then be filled before painting is begun. Because of the curing period ordinarily required for plastered walls, it is not advisable to apply oil-base paints until as least 60 days after plastering is completed. Water-mix or resin-base paints may be applied without necessity of an aging period.

Large plaster cracks often indicate a structural weakness in the framing. One of the common areas that may need correction is around a basement stairway. Framing may not be adequate for the loads of the walls and ceilings. In such cases, the use of an additional post and pedestal may be required to correct this fault. Inadequate framing around fireplaces and chimney openings, and joists that are not doubled under partitions, are other common sources of weakness.

Moisture on Windows

Points to be noted with respect to moisture on windows are as follows:

1. During cold weather, condensation, and in cold climates, frost will collect on the inner face of single-glazed windows. Water from the condensation or melting frost runs down the glass and soaks into the wood sash to cause stain, decay,

and paint failure. The water may rust steel sash. To prevent such condensation, the window should be provided with a storm sash. Double glazing will also minimize this condensation.

2. Occasionally, in very cold weather, frost may form on the inner surfaces of the storm windows. This may be caused by:

   (a) Loose-fitting window sash that allows moisture to enter the space between the window and storm sash.

   (b) High relative humidity in the living quarters.

Generally, the condensation on storm sash does not create a maintenance problem, but it may be a nuisance. Weatherstripping the inner sash offers increased resistance to moisture flow and may prevent this condensation. Lower relative humidities in the house are also helpful.

Moisture on Doors

Condensation may collect on single exterior doors during severe cold periods for the same reasons as described for windows. Here again, the water may cause damage to the door and to the finish. Storm doors offer the most practical means of preventing or minimizing such condensation. The addition of storm doors will also decrease the warping of the exterior doors.

FLOORS

A finish floor that has been improperly laid is a source of trouble for the housewife. This flooring may have been laid with varying moisture contents in the boards or at too high a moisture content. Cracks or openings in the floor appear during the heating season as the flooring dries out. If the floor has a few large cracks, one expedient is to fit matching strips of wood between the flooring strips and to glue them in place. In severe cases, it may be necessary to replace sections of the floor or to refloor the entire house. Another method would be to cover the existing flooring with a thin flooring, 5/16 or 3/8 inch thick. This would require removal of the base shoe, fitting the thin flooring around door jambs, and perhaps sawing off the door bottoms.

SUMMARY

A certain amount of foreseeing possible future trouble spots will often prevent a major repair bill. A damp basement will sometimes occur for several months after a house has been completed, and in most cases it will progressively disappear. In cases of persistent dampness, the owner should check various areas in order to eliminate this problem. Checking drainage downspouts for proper grade to carry away drain water is one of the first on the list. Some settling of the soil may occur at the foundation wall and form a water pocket.

If a house is designed with a crawl-space area, it should be checked for termite activity or damage. A well-constructed house will have a termite shield under the wood sill; this should also be checked for proper projection. Crawl-space ventilators should be checked for proper air circulation, as well as any obstruction which could prevent circulation.

The roof and attic should be inspected for humps in the asphalt-shingles which are an indication of loose nails. All flashing around chimneys, valleys, and ridges should be checked for possible leaks. Attics should be checked for proper ventilation and leaks in the roof.

Interior and exterior walls should be checked periodically for cracks, paint peeling, and air leaks. Windows should be checked for air leaks and freedom of operation. Floors should be checked for cracks or openings which appear during the heating season. Floor joists should be checked for cracks or decay.

REVIEW QUESTIONS

1. How often should a frame house be painted?
2. Why does plaster crack above windows and door openings?
3. How can dampness be corrected in a crawl-space?
4. What should be checked when inspecting a roof?
5. Why does a wood floor crack or buckle after being installed?

CHAPTER 22

Physical Characteristics of Wood

Wood, like all plant material, is made up of cells, or fibers, which when magnified have an appearance similar to, though less regular than, that of the common honeycomb. The walls of the honeycomb correspond to the walls of the fibers, and the cavities in the honeycomb correspond to the hollow or open spaces of the fibers.

SOFTWOODS AND HARDWOODS

All lumber is divided as a matter of convenience into two great groups, softwoods and hardwoods. The softwoods in general are the coniferous or cone-bearing trees, such as the various pines, spruces, hemlocks, firs, and cedar. The hardwoods are the noncone-bearing trees, such as the maple, oak, poplar, and the like. These terms are used as a matter of custom, for not all so-called softwoods are soft nor are all so-called hardwoods necessarily hard. As a matter of fact, such so-called softwoods as long-leaf southern pine, and Douglas fir are much harder than poplar, basswood, etc., which are called hardwoods.

Other and perhaps more accurate terms often used for these two groups are the needle-bearing trees and the broad-leaved trees, referring to the softwoods and hardwoods, respectively. In general, the softwoods are more commonly used for structural purposes such as for joists, studs, girders, posts, etc., while the hardwoods are more likely to be used for interior finish, flooring, and furniture. The softwoods are also used for interior finish and in many cases for floors, but are not often used for furniture.

319

MOISTURE CONTENT

While the tree is living, both the cells and cell walls are filled with water to an extent. As soon as the tree is cut, the water within the cells, or "free water" as it is called, begins to evaporate. This process continues until practically all of the "free water" has left the wood. When this stage is reached the wood is said to be at the fiber-saturation point; that is, what water is contained is mainly in the cell or fiber walls.

Except in a few species, there is no change in size during this preliminary drying process, and therefore no shrinkage during the evaporaton of the "free water." Shrinkage begins only when water begins to leave the cell walls themselves. What causes shrinkage and other changes in wood is not fully understood; but it is thought that as water leaves the cell walls, they contract, becoming harder and denser, thereby causing a general reduction in size of the piece of wood. If the specimen is placed in an oven which is maintained at 212° F., the temperature of boiling water, the water will evaporate and the specimen will continue to lose weight for a time. Finally a point is reached at which the weight remains substantially constant. This is another way of saying that all of the water in the cells and cell walls has been driven off. The piece is then said to be "oven dry."

If it is now taken out of the oven and allowed to remain in the open air, it will gradually take on weight, due to the absorption of moisture from the air. As when placed in the oven, a point is reached at which the weight of the wood in contact with the air remains more or less constant. Careful tests, however, show that it does not remain exactly constant, it will take on and give off water as the moisture in the atmosphere increases or decreases. Thus, a piece of wood will contain more water during the humid, moist summer months than in the colder, drier winter months. When the piece is in this condition it is in "equilibrium with the air" and is said to be "air-dry."

A piece of lumber cut from a green tree and left in the atmosphere in such a way that the air may circulate freely about it will gradually arrive at this air-dry condition. This ordinarily takes from one to three months, and the process is termed "air season-

ing." It can be greatly hastened by placing the wood in an artificial-
ly heated oven or "dry kiln" until the moisture content of the wood
is that of air dryness. The amount of water contained by wood in
the green condition varies greatly, not only with the species but
in the same species, and even in the same tree, according to the
position in the tree. But as a general average, at the fiber-saturation
point, most woods contain from 23 to 30 per cent of water as compared with the oven-dry weight of the wood. When air-dry, most
woods contain from 12 to 15 per cent of moisture.

SHRINKAGE

As the wood dries from the green state, which is that of the
freshly cut tree, to the fiber-saturation point, except in a few
species, there is no change other than that of weight. It has already
been pointed out that as the moisture dries out of the cell walls,
in addition to the decrease in weight, shrinkage results in a definite
decrease in size. It has been found, however, that there is little

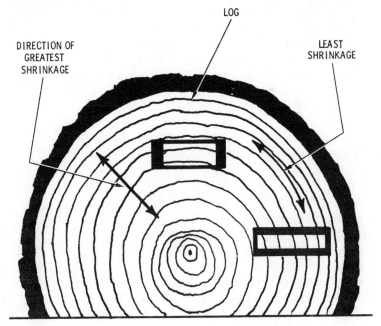

Fig. 1. Illustrating the shrinkage of wood from end to end.

or no decrease along the length of the grain, and that the decrease is at right angles to the grain.

This is an important consideration to be remembered when framing a building. For example, a stud in a wall will not shrink appreciably in length, whereas it will shrink somewhat in both the 2-inch and the 4-inch way. In like manner, a joist, if it is green when put in place, will change in depth as it seasons in the building. This shrinkage is illustrated in Fig. 1. These principles of shrinkage also explain why an edge-grain or quarter-sawed floor is less likely to open up than a flat-grain floor.

DENSITY

The tree undergoes a considerable impetus early every spring and grows very rapidly for a short time. Large amounts of water are carried through the cells to the rapidly growing branches and leaves at the top of the tree. This water passes upward mainly in the outer layers of the tree. The result is that the cells next to the bark, which are formed during the period of rapid growth, have thin walls and large passages. Later on, during the summer, the rate of growth slows up and the demand for water is less. The cells which are formed during the summer have much thicker walls and much smaller pores. Thus, a year's growth forms two types of wood—the spring wood, as it is called, being characterized by softness and openness of grain, and the summer wood by hardness and closeness, or density, of grain. The spring wood and summer wood growth for one year is called an "annual ring."

There is one ring for each year of growth. This development of spring wood and summer wood is a marked characteristic of practically all woods and is clearly evident in such trees as the yellow pines and firs, and less so in the white pines, maple, and the like. Careful examination will reveal this annual ring, however, in practically all species. It follows, therefore, that a tree in which the dense summer wood predominates is stronger than one in which the soft spring wood predominates. This is a point which should be borne in mind in selecting material for important members such as girders and posts carrying heavy loads. The strength of wood of the same species varies markedly with the density. For example, Douglas fir or southern pine, carefully selected for density, is one-sixth stronger than lumber of the same species and knot limitations in which the spring wood predominates. Trees having approximately one-third or more of cross-sectional area in summer wood fulfill one of the requirements for structural timbers.

ESTIMATING DENSITY

It must be remembered that the small cells or fibers which make up the wood structure are hollow. Wood substance itself has a specific gravity of about 1.5, and therefore will sink in water. It is stated that wood substance of all species is practically of the same density. Strength of wood depends upon its density and varies with its density. The actual dry weight of lumber is a good criterion of its strength, although weight can not always be relied upon as a basis for determining strength, as other important factors frequently must be considered in a specific piece of wood.

The hardness of wood is also another factor which assists in estimating the strength of wood. A test sometimes used is cutting across the grain. This test cannot be utilized in the commercial grading of lumber because a moisture content will affect the hardness and because hardness thus measured can not

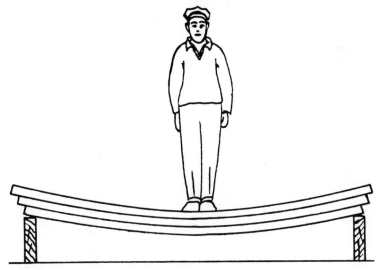

Fig. 2. Illustrating shear in lumber.

be adequately defined. The annual rings found in practically all species are an important consideration in estimating density, although the annual rings indicate different conditions in different species. In ring-porous hardwoods and in the conifers, where the contrast between spring wood and summer wood is definite, the proportion of hard summer wood is an indication of the strength of the individual piece of wood. The amount of summer wood, however, cannot always be relied upon as an indication of strength, because summer wood itself varies in density. When cut across the grain with a knife, the density of summer wood may be estimated on the basis of hardness, color, and luster.

In conifers, annual rings of average width indicate denser material or a larger proportion of summer wood than in wood with either wide or narrow rings. In some old conifers of virgin growth, in which the more recent annual rings are narrow, the wood is less dense than where there has been normal growth. On the other hand, in young trees where the growth has not been impeded by other trees, the rings are wider and in consequence the wood less dense. These facts may account for the belief that all second-growth timber and all sapwood are weak. In accepting wood for density, the contrast between summer wood and spring wood should be pronounced.

Oak, ash, hickory, and other ring-porous hardwoods in general rank high in strength when the annual rings are wide. In this respect they contrast with conifers. These species have more summer wood than spring wood as the rings become wider. For this reason, oak, hickory, ash, and elm of second growth are considered superior because of fast growth and increase in proportion of

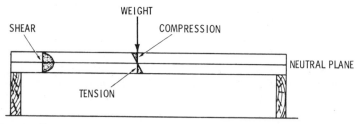

Fig. 3. Illustrating the different proportions of tension, compression, and shear.

summer wood. These conditions do not always exist, however, for exceptions occur, especially in ash and oak, where, although the summer wood is about normal, it may not be dense or strong. Very narrow rings in ring-porous hardwoods are likely to indicate weak and brashy material composed largely of spring wood with big pores. Maple, birch, beech, and other diffuse-porous hardwoods in general show no definite relationship between the width of rings and density, except that usually narrow rings indicate brash wood.

STRENGTH

Wood, when used in ordinary structures, is called upon to have three types of strength—tension, compression, and shear.

Tension

Tension is the technical term for a pulling stress. For example, if two men are having a tug of war with a rope, the rope is in tension. The tensile strength of wood, especially of the structural grades, is very high.

Compression

If, however, the men at opposite ends of a 2 by 4 are trying to push each other over, the timber is in compression. Tension and compression represent, therefore, exactly opposite forces.

Shear

Shear is harder to explain. If two or three planks are placed one upon the other between two blocks, and a person were to stand in the middle, the planks would bend and assume a position similar to that shown in Fig. 2. It will be noted that at the outer ends the boards tend to slip past each other.

If the planks were securely spiked through from top to bottom, the slipping would be in a great measure prevented and the boards would act more as one piece of wood. In very solid timber there is the same tendency for the various parts of the piece to slip past each other. This tendency is called *horizontal shear*. A defect, such as a check, which runs horizontally through a piece of a timber and tends to separate the upper from the lower part, is a weakness in shear.

It is well to analyze this matter a little further. Suppose that the planks were spiked through at the center of span only, i.e., half-way between the blocks. Such spikes would not increase the stiffness of the planks. It is clear, therefore that there is no horizontal shear near the center of the span (Fig. 3), and that the shear increases as one approaches either end of the beam. This will explain why, as most carpenters have doubtless observed, steel stirrups are used in concrete beams (weak in shear), why there is usually none near the center, and why they are put closer and closer together near the ends of the beams.

For all practical purposes the compressive strength of wood may be considered to equal its tensile strength. It has been extremely difficult to make any direct measurements of the tensile strength of wood. In an experiment designed to ascertain the tensile strength of a specimen of wood, a 4 by 4 inch piece was selected. A portion about a foot in length near the center was carefully cut down on all four sides until it was exactly 3/4 inch square. The test specimen was placed in a machine which gripped the 4 by 4 inch ends securely and a pull was exerted. The specimen did not pull apart. The 3/4-inch square section held and actually pulled out of the end of the 4 by 4, leaving a 3/4-inch square hole. This is an excellent illustration of how a piece may fail from shear rather than tension, the shear in this case being insufficient to prevent the 3/4-inch square piece from pulling out.

DEADWOOD

Because in some instances persons are prejudiced against the use of timber cut from dead trees, it is customary for individuals to specify that only timber cut from live trees will be accepted. It is true, however, that when sound trees which are dead are sawed into lumber and the weathered or charred outside is cut away, the resulting lumber can not be distinguished from that coming from live trees except in so far as the lumber from dead trees may be somewhat seasoned at the time it is sawed. It must be remembered that the heartwood of a living tree is fully matured and that in the sapwood only a small portion of the cells are in a living condition. As a consequence most of the wood cut from trees is already dead even when the tree itself is considered alive.

326

For structural purposes, it may be said that lumber cut from fire or insect killed trees is just as good as any other lumber unless the wood has been subjected to further decay or insect attack.

DRY-ROT

Loosely used, the term "dry-rot" is applied to any type of dry crumbly rot and includes under these circumstances all types of brown decay in wood. The pathologist uses the term "dry-rot" in the limited sense as it applies to the work of certain decay fungi which are frequently found growing in timber where they appear to have access to no moisture. Decay fungi will not grow in perfectly dry wood and no material decay need be expected in wood used under shelter and maintained in a normal air-dry condition. With moist wood the fungi are able to penetrate amazingly long distances because they extend their water-supply system by means of slender, minute, porous strands. Fungi of other kinds produce a rot not unlike dry-rot, but brown or yellow in color. The wood in an advanced state is shrunken, and in some places the cracks are filled with a white soggy mass, the wood itself being brittle, friable, and easily crushed into powder.

VIRGIN AND SECOND GROWTH

Occasionally an order calls for lumber of either virgin growth or second growth. The terms, however, are without significance, as an individual cannot tell one type from the other when it is delivered.

The virgin growth, which is also called old growth or first growth, refers to timber which grows in the forest along with many other trees, and therefore has suffered the consequence of the fight for sunlight and moisture.

The second growth is considered as that timber which grows up with less of the competition for sunlight and moisture which characterizes first-growth timber.

Because of environment, the virgin growth is usually thought of as wood of slow-growing type, whereas the second growth is considered as of relatively rapid growth, evidenced by wider annual rings. In such hardwoods as ash, hickory, elm, and oak these wider annual rings are supposed to indicate stronger and tougher wood, whereas in the conifers such as pine and fir, this condition is supposed to result in a weaker and brasher wood. For this reason, where the strength and toughness are desired, the second growth is preferred among hardwoods, and virgin growth is desired in conifers. Because of the variety of conditions under which both virgin and second growth timbers grow, because virgin growth may have the characteristics of second growth, and because second growth may have the characteristics of virgin timber, it is advisable in judging the strength of wood to rely upon its density and rate of growth rather than upon its being either virgin or second growth.

TIME OF CUTTING TIMBER

The time when timber is cut has very little to do with its durability or other desirable properties, if, after it is cut, it is cared for properly. Timber cut in the late spring, however, or early summer is more likely to be attacked by insects and fungi. In addition, seasoning will proceed much more rapidly during the summer months and, therefore, will result in checking, unless the lumber is shaded from the intense sunlight. It is stated that there is practically no difference in the moisture content in green lumber, either during the summer or winter.

AIR-DRIED AND KILN-DRIED WOOD

There is a prevailing misapprehension that air-dried lumber is stronger or better than kiln-dried lumber. Exhaustive tests have conclusively shown that good kiln-drying and good air-drying have exactly the same results upon the strength of the wood. Wood increases in strength with the elimination of moisture content. This may account for the claim that kiln-dried lumber is stronger than air-dried lumber. This has little significance, because in use wood will come to practically the same moisture content whether it has been kiln-dried or air-dried.

The same kiln-drying process cannot be applied to all species of wood. Consequently it must be remembered that lack of certain strength properties in wood may be due to improper kiln-drying.

Similar damage also may result from air seasoning under unsuitable conditions.

SAPWOOD VERSUS HEARTWOOD

The belief is common that in some species the heartwood is stronger than the sapwood and that the reverse is the case in such species as hickory and ash. Tests have shown conclusively that neither is the case, and that sapwood is not necessarily stronger than heartwood or heartwood stronger than sapwood, but that density rather than other factors makes the difference in strength. In trees that are mature, the sapwood is frequently weaker, whereas in young trees the sapwood may be stronger. Density, proportion of spring and summer wood, then must be the basis of consideration of strength rather than whether the wood is sapwood or heartwood. Under unfavorable conditions, the sapwood of most species is more subject to decay than the heartwood.

BLUE STAIN

In the sapwood of many species of both softwoods and hardwoods, there often develops a bluish-black discoloration known as blue stain. It does not indicate an early stage of decay, nor does it have any practicable effect on the strength of the wood. Blue stain is caused by a fungus growth in unseasoned lumber. Although objectionable where appearance is of importance, as in unpainted sash or trim, blue stain need cause no concern for framing lumber. Precautions should be taken, however, to make sure that no decay fungus is present with the blue stain.

SUMMARY

Lumber is divided as a matter of convenience into two groups—softwood and hardwood. Softwoods are generally cone-bearing trees such as various pines, spruces, hemlocks, firs, and cedars. Hardwoods are noncone-bearing trees such as maple, oak, and poplar. These terms are used only as a classification and do not mean that so-called softwood is particularly soft, nor that hardwood is particularly hard.

Except for a few species, there is no change in size during the preliminary drying process, and therefore no shrinkage during the evaporation of the free water. There is generally no change in the lumber other than that of weight. There is little or no decrease along the length of the grain—the decrease, if any, will be at right angles to the grain. If green or wet lumber is used in construction, the length will not decrease—it will be the depth and width that will change.

When using wood in ordinary structures, it is called upon to have three types of strength—tension, shear, and compression. Tension is the pulling strength of wood; shear is the strength available at both ends of a piece of timber. Compression is the opposite of tension, whereas tension is the pulling force and compression is the pushing force.

REVIEW QUESTIONS

1. What are tension, compression, and shear of wood?
2. Explain the moisture content of lumber.
3. What is the percentage of shrinkage in drying lumber?
4. What is dry-rot?
5. Explain blue stain in lumber.

Glossary of Housing Terms

Air-Dried Lumber—Lumber that has been piled in yards or sheds for any length of time. For the United States as a whole, the minimum moisture content of thoroughly air-dried lumber is 12 to 15 percent, with the average somewhat higher.

Airway—A space between roof insulation and roof boards for movement of air.

Alligatoring—Coarse checking pattern characterized by a slipping of the new paint coating over the old coating to the extent that the old coating can be seen through the fissures.

Anchor Bolts—Bolts to secure a wooden sill to a concrete or masonry floor, foundation, or wall.

Apron—The flat member of the inside trim of a window placed against the wall immediately beneath the stool.

Areaway—An open subsurface space adjacent to a building used to admit light or air, or as a means of access to a basement or cellar.

Asphalt—Most native asphalt is a residue from evaporated petroleum. It is insoluble in water but is soluble in gasoline and melts when heated. Used widely in building for waterproofing roof coverings of many types, exterior wall coverings, flooring tile, and the like.

Astragal—A moulding, attached to one of a pair of swinging doors, against which the other door strikes.

Attic Ventilators—In home building, usually openings in gables or ventilators in the roof. Also, mechanical devices to force ventilation by the use of power-driven fans. Also see Louver.

Backband—Moulding used on the side of a door or window casing for ornamentation or to increase the width of the trim.

Backfill—The replacement of excavated earth into a pit or trench or against a foundation wall.

Balusters—Small spindles or members forming the main part of a railing for a stairway or balcony, fastened between a bottom and top rail.

Base or Baseboard—A board placed against the wall around a room next to the floor to provide a proper finish between the floor and plaster.

Base Moulding—Moulding used to trim the upper edge of interior baseboard.

Base Shoe—Moulding used next to the floor on interior baseboard. Sometimes called a carpet strip.

Batten—Narrow strips of wood or metal used to cover joints.

Batter Board—One of a pair of horizontal boards nailed to posts set at the corners of an excavation, used to indicate the desired level. Also a fastening for stretched strings to indicate the outlines of foundation walls.

Bay Window—Any window space projecting outward from the walls of a building, either square or polygonal in plan.

Beam—A structural member transversely supporting a load.

Bearing Partition—A partition that supports any vertical load in addition to its own weight.

Bearing Wall—A wall that supports any vertical load in addition to its own weight.

Bed Moulding—A moulding in an angle, as between an overhanging cornice or eaves, of a building and the side walls.

Blinds (Shutters)—Light wood sections in the form of doors to close over windows to shut out light, give protection, or add temporary insulation. Commonly used now for ornamental purposes, in which case they are fastened rigidly to the building.

Blind-Nailing—Nailing in such a way that the nailheads are not visible on the face of the work.

Blind Stop—A rectangular moulding, usually 3/4 by 1-3/8 inches or more, used in the assembly of a window frame.

Blue Stain—A bluish or grayish discoloration of the sapwood caused by the growth of certain moldlike fungi on the surface and in the interior of the piece, made possible by the same conditions that favor the growth of other fungi.

Bodied Linseed Oil—Linseed oil that has been thickened in viscosity by suitable processing with heat or chemicals. Bodied oils are obtainable in a great range in viscosity from a little greater than that of raw oil to just short of a jellied condition.

Boiled Linseed Oil—Linseed oil in which enough lead, manganese, or cobalt salts have been incorporated to make the oil harden more rapidly when spread in thin coatings.

Bolster—A short horizontal timber resting on the top of a column for the support of beams or girders.

Boston Ridge—A method of applying asphalt or wood shingles as a finish at the ridge or hips of a roof.

Brace—An inclined piece of framing lumber used to complete a triangle, and thereby to stiffen a structure.

Brick Veneer—A facing of brick laid against frame or tile wall construction.

Bridging—Small wood or metal members that are inserted in a diagonal position between the floor joists to act both as tension and compression members for the purpose of bracing the joists and spreading the action of loads.

Buck—Often used in reference to rough frame opening members. Door bucks used in reference to metal door frame.

Built-up Roof—A roofing composed of three to five layers of rag felt or jute saturated with coal tar, pitch, or asphalt. The top is finished with crushed slag or gravel. Generally used on flat or low-pitched roofs.

Butt Joint—The junction where the ends of two timbers or other members meet in a square-cut joint.

Cabinet—A shop- or job-built unit for kitchens. Cabinets often include combinations of drawers, doors, and the like.

Cant Strip—A wedge or triangular-shaped piece of lumber used at the gable ends under the shingles or at the junction of the house and a flat deck under the roofing.

Cap—The upper member of a column, pilaster, door cornice, moulding, and the like.

Casing—Wide moulding of various widths and thicknesses used to trim door and window openings.

Casement Frames and Sash—Frames of wood or metal enclosing part or all of the sash which may be opened by means of hinges affixed to the vertical edges.

Cement, Keene's—The whitest finish plaster obtainable that produces a wall of extreme durability. Because of its density, it excels for a wainscoting plaster for bathrooms and kitchens and is also used extensively for the finish coat in auditoriums, public buildings, and other places where walls will be subjected to unusually hard wear or abuse.

Checking—Fissures that appear with age in many exterior paint coatings, at first superficial, but which in time may penetrate entirely through the coating.

Checkrails—Meeting rails sufficiently thicker than a window to fill the opening between the top and bottom sash made by the parting stop in the frame. They are usually beveled.

Collar Beam—A beam connecting pairs of opposite rafters above the attic floor.

Column—In architecture: A perpendicular supporting member, circular or rectangular in section, usually consisting of a base, shaft, and capital. In engineering: A structural compression member, usually vertical, supporting loads acting on or near and in the direction of its longitudinal axis.

Combination Doors—Combination doors have an inside removable section so that the same frame serves for both summer and

334

winter protective devices. A screen is inserted in warm weather to make a screen door, and a glazed or a glazed-and-wood-paneled section in winter to make a storm door. The inconvenience of handling a different door in each season is eliminated.

Concrete, Plain—Concrete without reinforcement, or reinforced only for shrinkage or temperature changes.

Condensation—Beads or drops of water, and frequently frost in extremely cold weather, that accumulate on the inside of the exterior covering of a building when warm, moisture-laden air from the interior reaches a point where the temperature no longer permits the air to sustain the moisture it holds. Use of louvers or attic ventilators will reduce moisture condensation in attics.

Conduit, Electrical—A pipe, usually metal, in which wire is installed.

Construction, Dry-Wall—A type of construction in which the interior wall finish is applied in a dry condition, generally in the form of sheet materials, as contrasted to plaster.

Construction, Frame—A type of construction in which the structural parts are of wood or dependent upon a wood frame for support. In codes, if brick or other incombustible material is applied to the exterior walls, the classification of this type of construction is usually unchanged.

Coped Joint—See Scribing.

Corbel Out—To build out one or more courses of brick or stone from the face of a wall, to form a support for timbers.

Corner Bead—A strip of formed galvanized iron, sometimes combined with a strip of metal lath, placed on corners before plastering to reinforce them. Also, a strip of wood finish three-quarters-round or angular placed over a plastered corner for protection.

Corner Boards—Used as trim for the external corners of a house or other frame structure against which the ends of the siding are finished.

Corner Braces—Diagonal braces let into studs to reinforce corners of frame structures.

Cornerite—Metal-mesh lath cut into strips and bent to a right angle. Used in interior corners of walls and ceilings on lath to prevent cracks in plastering.

Cornice—A decorative element made up of moulded members usually placed at or near the top of an exterior or interior wall.

Cornice Return—That portion of the cornice that returns on the gable end of a house.

Counterflashing—A flashing usually used on chimneys at the roofline to cover shingle flashing and to prevent moisture entry.

Cove Moulding—A three-sided moulding with concave face used wherever small angles are to be covered.

Crawl Space—A shallow space below the living quarters of a house. It is generally not excavated or paved and is often enclosed for appearance by a skirting or facing material.

Cricket—A small drainage diverting roof structure of single or double slope placed at the junction of larger surfaces that meet at an angle.

Crown Moulding—A moulding used on a cornice or wherever a large angle is to be covered.

d—See Penny.

Dado—A rectangular groove in a board or plank. In interior decoration, a special type of wall treatment.

Decay—Disintegration of wood or other substance through the action of fungi.

Deck paint—An enamel with a high degree of resistance to mechanical wear, for use on such surfaces as porch floors.

Density—The mass of substance in a unit volume. When expressed in the metric system, it is numerically equal to the specific gravity of the same substance.

Dimension—See Lumber, dimension.

Direct Nailing—To nail perpendicular to the initial surface or to the junction of the pieces joined. Also termed *face nailing.*

Doorjamb, Interior—The surrounding case into which and out of which a door closes and opens. It consists of two upright pieces, called jambs, and a head, fitted together and rabbeted.

Dormer—An interal recess, the framing of which projects from a sloping roof.

Downspout—A pipe, usually of metal, for carrying rainwater from roof gutters.

Dressed and Matched (Tongue and Groove)—Boards or planks machined in such a manner that there is a groove on one edge and a corresponding tongue on the other.

Drier, Paint—Usually oil-soluble soaps of such metals as lead, manganese, or cobalt, which, in small proportions, hasten the oxidation and hardening (drying) of the drying oils in paints.

Drip—(*a*) A member of a cornice or other horizontal exterior-finish course that has a projection beyond the other parts for throwing off water. (*b*) A groove in the under side of a sill to cause water to drop off on the outer edge, instead of drawing back and running down the face of the building.

Drip Cap—A moulding placed on the exterior top side of a door or window to cause water to drip beyond the outside of the frame.

Ducts—In a house, usually round or rectangular metal pipes for distributing warm air from the heating plant to rooms, or air from a conditioning device. Ducts are also made of asbestos and composition materials.

Eaves—The margin or lower part of a roof projecting over the wall.

Expansion Joint—A bituminous fiber strip used to separate blocks or units of concrete to prevent cracking due to expansion as a result of temperature changes.

Facia or Fascia—A flat board, band, or face, used sometimes by itself but usually in combination with mouldings, often located at the outer face of the cornice.

Filler (Wood)—A heavily pigmented preparation used for filling and leveling off the pores in open-pored woods.

Fire-Resistive—In the absence of a specific ruling by the authority having jurisdiction, applies to materials for construction not combustible in the temperatures of ordinary fires and that will withstand such fires without serious impairment of their usefulness for at least 1 hour.

Fire-Retardant Chemical—A chemical or preparation of chemicals used to reduce flammability or to retard spread of flame.

Fire Stop—A solid, tight closure of a concealed space, placed to prevent the spread of fire and smoke through such a space.

Flagstone (Flagging or Flags)—Flat stones, from 1 to 4 inches thick, used for rustic walks, steps, floors, and the like. Usually sold by the ton.

Flashing—Sheet metal or other material used in roof and wall construction to protect a building from seepage of water.

Flat Paint—An interior paint that contains a high proportion of pigment, and dries to a flat or lusterless finish.

Flue—The space or passage in a chimney through which smoke, gas, or fumes ascend. Each passage is called a flue, which, together with any others and the surrounding masonry, make up the chimney.

Flue Lining—Fire clay or terra-cotta pipe, round or square, usually made in all of the ordinary flue sizes and in 2-foot lengths, used for the inner lining of chimneys with the brick or masonry work around the outside. Flue lining should run from the concrete footing to the top of the chimney cap. Figure a foot of flue lining for each foot of chimney.

Footing—The spreading course or courses at the base or bottom of a foundation wall, pier, or column.

Foundation—The supporting portion of a structure below the first-floor construction, or below grade, including the footings.

338

Framing, Balloon—A system of framing a building in which all vertical structural elements of the bearing walls and partitions consist of single pieces extending from the top of the soleplate to the roofplate and to which all floor joists are fastened.

Framing, Platform—A system of framing a building in which floor joists of each story rest on the top plates of the story below or on the foundation sill for the first story, and the bearing walls and partitions rest on the subfloor of each story.

Frieze—Any sculptured or ornamental band in a building. Also the horizontal member of a cornice set vertically against the wall.

Frostline—The depth of frost penetration in soil. This depth varies in different parts of the country. Footings should be placed below this depth to prevent movement.

Fungi, Wood—Microscopic plants that live in damp wood and cause mold, stain, and decay.

Fungicide—A chemical that is poisonous to fungi.

Furring—Strips of wood or metal applied to a wall or other surface to even it, to form an air space, or to give the wall an appearance of greater thickness.

Gable—That portion of a wall contained between the slope of a single-sloped roof and a line projected horizontally through the lowest elevation of the roof construction.

Gable End—An end wall having a gable.

Gloss Enamel—A finishing material made of varnish and sufficient pigments to provide opacity and color, but little or no pigment of low opacity. Such an enamel forms a hard coating that has a maximum smoothness of surface and a high degree of gloss.

Gloss (Paint or Enamel)—A paint or enamel that contains a relatively low proportion of pigment and dries to a sheen or luster.

Girder—A large or principal beam used to support concentrated loads at isolated points along its length.

Grain—The direction, size, arrangement, appearance, or quality of the fibers in wood.

Grain, Edge (Vertical)—Edge-grain lumber has been sawed parallel to the pith of the log and approximately at right angles to the growth rings; i.e., the rings form an angle of 45° or more with the surface of the piece.

Grain, Flat—Flat-grain lumber has been sawed parallel to the pith of the log and approximately at right angles to the growth rings; i.e., the rings form an angle of less than 45° with the surface of the piece.

Grain, Quartersawn—Another term for edge grain.

Grounds—Strips of wood, of the same thickness as the lath and plaster, that are attached to walls before the plastering is done. Used around windows, doors, and other openings as a plaster stop and in other places for the purpose of attaching baseboards or other trim.

Grout—Mortar made of such consistency by the addition of water that it will just flow into the joints and cavities of the masonry work and fill them solid.

Gutter or Eave Trough—A shallow channel or conduit of metal or wood set below and along the eaves of a house to catch and carry off rainwater from the roof.

Gypsum Plaster—Gypsum formulated to be used with the addition of sand and water for base-coat plaster.

Header—(*a*) A beam placed perpendicular to joists and to which joists are nailed in framing for chimney, stairway, or other opening. (*b*) A wood lintel.

Hearth—The floor of a fireplace, usually made of brick, tile, or stone.

Heartwood—The wood extending from the pith to the sapwood, the cells of which no longer participate in the life processes of the tree.

Hip—The external angle formed by the meeting of two sloping sides of a roof.

Hip Roof—A roof that rises by inclined planes from all four sides of a building.

Humidifier—A device designed to discharge water vapor into a confined space for the purpose of increasing or maintaining the relative humidity in an enclosure.

I-Beam—A steel beam with a cross section resembling the letter "I."

Insulating Board or Fiberboard—A low-density board made of wood, sugarcane, cornstalks, or similar materials, usually formed by a felting process, dried and usually pressed to thicknesses of 1/2 and 25/32 inch.

Insulation, Building—Any material high in resistance to heat transmission that, when placed in the walls, ceilings, or floor of a structure, will reduce the rate of heat flow.

Jack Rafter—A rafter that spans the distance from the wallplate to a hip, or from a valley to a ridge.

Jamb—The side post or lining of a doorway, window, or other opening.

Joint—The space between the adjacent surfaces of two members or components joined and held together by nails, glue, cement, mortar, or other means.

Joint Cement—A powder that is usually mixed with water and used for joint treatment in gypsum-wallboard finish. Often called "spackle."

Joist—One of a series of parallel beams used to support floor and ceiling loads, and supported in turn by larger beams, girders, or bearing walls.

Knot—That portion of a branch or limb that has become incorporated in the body of a tree.

Landing—A platform between flights of stairs or at the termination of a flight of stairs.

Lath—A building material of wood, metal, gypsum, or insulating board that is fastened to the frame of a building to act as a plaster base.

Lattice—An assemblage of wood or metal strips, rods, or bars made by crossing them to form a network.

Leader—See Downspout.

Ledger Strip—A strip of lumber nailed along the bottom of the side of a girder on which joists rest.

Light—Space in a window sash for a single pane of glass. Also, a pane of glass.

Lintel—A horizontal structural member that supports the load over an opening such as a door or window.

Lookout—A short wood bracket or cantilever to support an over-hanging portion of a roof or the like, usually concealed from view.

Louver—An opening with a series of horizontal slats so arranged as to permit ventilation but to exclude rain, sunlight, or vision. See also Attic Ventilators.

Lumber—Lumber is the product of the sawmill and planing mill not further manufactured other than by sawing, resawing, and passing lengthwise through a standard planing machine, cross-cut to length, and matched.

Lumber, Boards—Yard lumber less than 2 inches thick and 2 or more inches wide.

Lumber, Dimension—Yard lumber from 2 inches to, but not including, 5 inches thick, and 2 or more inches wide. Includes joists, rafters, studding, planks, and small timbers.

Lumber, Dressed Size—The dimensions of lumber after shrinking from the green dimension and after planing, usually 3/8 inch less than the nominal or rough size. For example, a 2-by-4 stud actually measures 1-5/8 by 3-5/8 inches.

Lumber, Matched—Lumber that is edge-dressed and shaped to

342

make a close tongue-and-groove joint at the edges or ends when laid edge to edge or end to end.

Lumber, Shiplap—Lumber that is edge-dressed to make a close rabbeted or lapped joint.

Lumber, Timbers—Yard lumber 5 or more inches in the least dimension. Includes beams, stringers, posts, caps, sills, girders, and purlins.

Lumber, Yard—Lumber of those grades, sizes, and patterns which are generally intended for ordinary construction, such as framework and rough coverage of houses.

Mantel—The shelf above a fireplace. Originally referred to the beam or lintel supporting the arch above the fireplace opening. Used also in referring to the entire finish around a fireplace, covering the chimney breast across the front and sometimes on the sides.

Masonry—Stone, brick, concrete, hollow-tile, concrete-block, gypsum-block, or other similar building units or materials or a combination of the same, bonded together with mortar to form a wall, pier, buttress, or similar mass.

Metal Lath—Sheets of metal that are slit and drawn out to form openings on which plaster is spread.

Millwork—Generally all building materials made of finished wood and manufactured in millwork plants and planing mills are included under the term *millwork*. It includes such items as inside and outside doors, window and doorframes, blinds, porchwork, mantels, panelwork, stairways, mouldings, and interior trim. It does not include flooring, ceiling, or siding.

Miter—The joining of two pieces at an angle that bisects the angle of junction.

Moisture Content of Wood—Weight of the water contained in the wood, usually expressed as a percentage of the weight of the oven-dry wood.

Mortise—A slot cut into a board, plank, or timber, usually edgewise, to receive the tenon of another board, plank, or timber to form a joint.

Moulding—Material, usually patterned strips, used to provide ornamental variation of outline or contour, whether projections or cavities, such as cornices, bases, window and doorjambs, and heads.

Mullion—A slender bar or pier forming a division between panels or units of windows, screens, or similar frames.

Muntin—The members dividing the glass or openings of sash, doors, and the like.

Natural Finish—A transparent finish, usually a drying oil, sealer, or varnish, applied on wood for the purpose of protection against soiling or weathering. Such a finish may not seriously alter the original color of the wood or obscure its grain pattern.

Newel—Any post to which a stair railing or balustrade is fastened.

Nonbearing Wall—A wall supporting no load other than its own weight.

Nosing—The projecting edge of a moulding or drip. Usually applied to the projecting moulding on the edge of a stair tread.

O. C. (On Center)—The measurement of spacing for studs, rafters, joists, and the like in a building from center of one member to the center of the next member.

O. G. (Ogee)—A moulding with a profile in the form of a letter S; having the outline of a reversed curve.

Paint—L, pure white lead (basic-carbonate) paint; TLZ, titanium-lead-zinc paint; TZ, titanium-zinc paint.

Panel—A large, thin board or sheet of lumber, plywood, or other material. A thin board with all its edges inserted in a groove of a surrounding frame of thick material. A portion of a flat surface recessed or sunk below the surrounding area, distinctly set off by moulding or some other decorative device. Also, a section of floor, wall, ceiling, or roof, usually prefabricated and of large size, handled as a single unit in the operations of assembly and erection.

Paper, Building—A general term for papers, felts, and similar sheet materials used in buildings without reference to their properties or uses.

Paper, Sheathing—A building material, generally paper or felt, used in wall and roof construction as a protection against the passage of air and sometimes moisture.

Parting Stop or Strip—A small wood piece used in the side and head jambs of double-hung windows to separate the upper and lower sash.

Partition—A wall that subdivides spaces within any story of a building.

Penny—As applied to nails, it originally indicated the price per hundred. The term now serves as a measure of nail length and is abbreviated by the letter *d*.

Pier—A column of masonry, usually rectangular in horizontal cross section, used to support other structural members.

Pigment—A powdered solid in suitable degree of subdivision for use in paint or enamel.

Pitch—The incline or rise of a roof. Pitch is expressed in inches or rise per foot of run, or by the ratio of the rise to the span.

Pitch Pocket—An opening extending parallel to the annual rings of growth, that usually contains, or has contained, either solid or liquid pitch.

Pith—The small, soft core at the original center of a tree around which wood formation takes place.

Plate—(*a*) A horizontal structural member placed on a wall or supported on posts, studs, or corbels to carry the trusses of a roof or to carry the rafters directly. (*b*) A shoe, or base member, as of a partition or other frame. (*c*) A small, relatively flat member placed on or in a wall to support girders, rafters, etc.

Plough—To cut a groove, as in a plank.

Plumb—Exactly perpendicular; vertical.

Ply—A term to denote the number of thicknesses or layers of roofing felt, veneer in plywood, or layers in built-up materials, in any finished piece of such material.

Plywood—A piece of wood made of three or more layers of veneer joined with glue and usually laid with the grain of adjoining plies at right angles. Almost always an odd number of plies are used to provide balanced construction.

Porch—A floor extending beyond the exterior walls of a building. It may be covered and enclosed or unenclosed.

Pores—Wood cells of comparatively large diameter that have open ends and are set one above the other to form continuous tubes. The openings of the vessels on the surface of a piece of wood are referred to as pores.

Preservative—Any substance that, for a reasonable length of time, will prevent the action of wood-destroying fungi, borers of various kinds, and similar destructive life when the wood has been properly coated or impregnated with it.

Primer—The first coat of paint in a paint job that consists of two or more coats; also the paint used for such a first coat.

Putty—A type of cement usually made of whiting and boiled linseed oil, beaten or kneaded to the consistency of dough and used in sealing glass in sash, filling small holes and crevices in wood, and for similar purposes.

Quarter Round—A moulding that presents a profile of a quarter circle.

Rabbet—A rectangular longitudinal groove cut in the corner of a board or other piece of material.

Radiant Heating—A method of heating, usually consisting of coils or pipes placed in the floor, wall, or ceiling.

Rafter—One of a series of structural members of a roof designed to support roof loads. The rafters of a flat roof are sometimes called roof joists.

Rafter, Hip—A rafter that forms the intersection of an external roof angle.

Rafter, Jack—A rafter that spans the distance from a wallplate to a hip or from a valley to a ridge.

Rafter, Valley—A rafter that forms the intersection of an internal roof angle.

Rail—A horizontal bar or timber of wood or metal extending from one post or support to another as a guard or barrier in a fence, balustrade, staircase, etc. Also, the cross or horizontal members of the framework of a sash, door, blind, or any paneled assembly.

Rake—The trim members that run parallel to the roof slope and from the finish between wall and roof.

Raw Linseed Oil—The crude product expressed from flaxseed and usually without much subsequent treatment.

Reflective Insulation—Sheet material with one or both surfaces of comparatively low heat emissivity that, when used in building construction so that the surfaces face air space, reduces the radiation across the air space.

Reinforcing—Steel rods or metal fabric placed in concrete slabs, beams, or columns to increase their strength.

Resin-Emulsion Paint—Paint, the vehicle (liquid part) of which consists of resin or varnish dispersed in fine droplets in water, analogous to cream (which is butterfat dispersed in water).

Relative Humidity—The amount of water vapor expressed as a percentage of the maximum quantity that could be present in the atmosphere at a given temperature. (The actual amount of water vapor that can be held in space increases with the temperature.)

Ribbon—A narrow board let into the studding to add support to joists.

Ridge—The horizontal line at the junction of the top edges of two sloping roof surfaces. The rafters are nailed at the ridge.

Ridge Board—The board placed on edge at the ridge of the roof to support the upper ends of the rafters.

Rise—The height a roof rising in horizontal distance (run) from the outside face of a wall supporting the rafters or trusses to the ridge of the roof. In stairs, the perpendicular height of a step or flight of steps.

Riser—Each of the vertical boards closing the spaces between the treads of the stairways.

Roll Roofing—Roofing material, composed of fiber and saturated with asphalt, that is supplied in rolls containing 108 square feet in 36-inch widths. It is generally furnished in weights of 55 to 90 pounds per roll.

Roof Sheathing—The boards or sheet material fastened to the roof rafters on which the shingles or other roof covering is laid.

Rubber-Emulsion Paint—Paint, the vehicle of which consists of rubber or synthetic rubber dispersed in fine droplets in water.

Run—In reference to roofs, the horizontal distance from the face of a wall to the ridge of the roof. Referring to stairways, the net width of a step; also the horizontal distance covered by a flight of steps.

Sapwood—The outer zone of wood, next to the bark. In the living tree it contains some living cells (the heartwood contains none), as well as dead and dying cells. In most species, it is lighter colored than the heartwood. In all species, it is lacking in decay resistance.

Sash—A single frame containing one or more panes of glass.

Sash Balance—A device, usually operated with a spring, designed to counter-balance window sash. Use of sash balances eliminates the need for sash weights, pulleys, and sash cord.

Saturated Felt—A felt which is impregnated with tar or asphalt.

Scratch Coat—The first coat of plaster, which is scratched to form a bond for the second coat.

Scribing—Fitting woodwork to an irregular surface.

Sealer—A finishing material, either clear or pigmented, that is usually applied directly over uncoated wood for the purpose of sealing the surface.

Seasoning—Removing moisture from green wood in order to improve its serviceability.

Semigloss Paint or Enamel—A paint or enamel made with a slight insufficiency of nonvolatile vehicle so that its coating, when dry, has some luster but is not very glossy.

Shake—A handsplit shingle, usually edge grained.

Sheathing—The structural covering, usually wood boards, plywood, or wallboards, placed over exterior studding or rafters of a structure.

Sheathing Paper—See Paper, Sheathing.

Shellac—A transparent coating made by dissolving lac, a resinous secretion of the lac bug (a scale insect that thrives in tropical countries, especially India), in alcohol.

Shingles—Roof covering of asphalt, asbestos, wood, tile, slate, or other material cut to stock lengths, widths, and thicknesses.

Shingles, Siding—Various kinds of shingles, some especially designed, that can be used as the exterior side-wall covering for a structure.

Shiplap—See Lumber, Shiplap.

Siding—The finish covering of the outside wall of a frame building, whether made of weatherboards, vertical boards with battens, shingles, or other material.

Siding, Bevel (Lap Siding)—Used as the finish siding on the exterior of a house or other structure. It is usually manufactured by resawing dry square-surfaced boards diagonally to produce two wedge-shaped pieces. These pieces commonly run from 3/16 inch thick on the thin edge to 1/2 to 3/4 inch thick on the other edge, depending on the width of the siding.

Siding, Drop—Usually 3/4 inch thick and 6 inches wide, machined into various patterns. Drop siding has tongue-and-groove joints, is heavier, has more structural strength, and is frequently used on buildings that require no sheathing, such as garages and barns.

Sill—The lowest member of the frame of a structure, resting on

349

the foundation and supporting the uprights of the frame. The member forming the lower side of an opening, as a door sill, window sill, etc.

Soffit—The underside of the members of a building, such as staircases, cornices, beams, and arches, relatively minor in area as compared with ceilings.

Soil Cover (Ground Cover)—A light roll roofing or plastic used on the ground of crawl spaces to minimize moisture permeation of the area.

Soil Stack—A general term for the vertical main of a system of soil, waste, or vent piping.

Sole or Soleplate—A member, usually a 2-by-4, on which wall and partition studs rest.

Span—The distance between structural supports, such as walls, columns, piers, beams, girders, and trusses.

Splash Block—A small masonry block laid with the top close to the ground surface to receive roof drainage and to carry it away from the building.

Square—A unit of measure—100 square feet—usually applied to roofing material. Side-wall coverings are often packed to cover 100 square feet and are sold on that basis.

Stain, Shingle—A form of oil paint, very thin in consistency, intended for coloring wood with rough surfaces, like shingles, but without forming a coating of significant thickness or with any gloss.

Stair Landing—A platform between flights of stairs or at the termination of a flight of stairs.

Stair Rise—The vertical distance from the top of one stair tread to the top of the one next above.

Stair Carriage—A stringer for steps on stairs.

Stool—The flat, narrow shelf forming the top member of the interior trim at the bottom of a window.

Storm Sash or Storm Window—An extra window usually placed on the outside of an existing window as additional protection against cold weather.

Story—That part of a building between floors and the floor or roof above.

String, Stringer—A timber or other support for cross members. In stairs, the support on which the staair treads rest; also *stringboard*.

Stucco—Most commonly refers to an outside plaster made with Portland cement as its base.

Stud—One of a series of slender wood or metal structural members placed as supporting eiements in walls and partitions. (Plural: studs or studding.)

Subfloor—Boards or sheet material laid on joists over which a finish floor is to be laid.

Tail Beam—A relatively short beam or joist supported in a wall on one end and by a header on the other.

Trimmer—A beam or joist to which a header is nailed in framing for a chimney, stairway, or other opening.

Termite—Insects that superficially resemble ants in size, general appearance, and habit of living in colonies; hence, they are frequently called *white ants*. Subterranean termites do not establish themselves in buildings by being carried in with lumber but by entering from ground nests after the building has been constructed. If unmolested, they eat out the woodwork, leaving a shell of sound wood to conceal their activities, and damage may proceed so far as to cause collapse of parts of a structure before discovery. There are about 56 species of termites known in the United States; but the two major species, classified from the manner in which they attack wood, are ground-inhabiting or subterranean termites, the most common, and dry-wood termites, found almost exclusively along the extreme southern border and the Gulf of Mexico in the United States.

Termite Shield—A shield, usually of noncorrodible metal, placed in or on a foundation wall or other mass of masonry or around pipes to prevent passage of termites.

351

Threshold—A strip of wood or metal beveled on each edge and used above the finished floor under outside doors.

Toenailing—To drive a nail at a slant with the initial surface in order to permit it to penetrate into a second member.

Tread—The horizontal board in a stairway on which the foot is placed.

Truss—A frame or jointed structure designed to act as a beam of long span, while each member is usually subjected to longitudinal stress only, either tension or compression.

Trim—The finish materials in a building, such as mouldings, applied around openings (window trim, door trim) or at the floor and ceiling of rooms (baseboard, cornice, picture moulding).

Turpentine—A volatile oil used as a thinner in paints and as a solvent in varnishes. Chemically, it is a mixture of terpenes.

Undercoat—A coating applied prior to the finishing or top coats of a paint job. It may be the first of two or the second of three coats. In some usage of the word it may become synonymous with priming coat.

Valley—The internal angle formed by the junction of two sloping sides of a roof.

Vapor Barrier—Material used to retard the flow of vapor or moisture into walls and thus to prevent condensation within them. There are two types of vapor barriers, the membrane that comes in rolls and is applied as a unit in the wall or ceiling construction, and the paint type, which is applied with a brush. The vapor barrier must be a part of the warm side of the wall.

Varnish—A thickened preparation of drying oil, or drying oil and resin, suitable for spreading on surfaces to form continuous, transparent coatings, or for mixing with pigments to make enamels.

Vehicle—A liquid portion of a finishing material; it consists of the binder (nonvolatile) and volatile thinners.

Veneer—Thin sheets of wood.

Vent—A pipe installed to provide a flow of air to or from a drainage system or to provide a circulation of air within such systems to protect trap seals from siphonage and back pressure.

Vermiculite—A mineral closely related to mica, with the faculty of expanding on heating to form lightweight material with insulation quality. Used as bulk insulation and also as aggregate in insulating and acoustical plaster and in insulating concrete floors.

Volatile Thinner—A liquid that evaporates readily and is used to thin or reduce the consistency of finishes without altering the relative volumes of pigments and nonvolatile vehicle.

Wallboard—Woodpulp, gypsum, or other materials made into large rigid sheets that may be fastened to the frame of a building to provide a surface finish.

Wane—Bark, or lack of wood or bark from any cause, on the edge or corner of a piece.

Wash—The upper surface of a member or material when given a slope to shed water.

Water Repellent—A liquid designed to penetrate into wood and to impart water repellancy to the wood.

Water Table—A ledge or offset on or above a foundation wall, for the purpose of shedding water.

Weatherstrip—Narrow strips made of metal, or other material, so designed that when installed at doors or windows they will retard the passage of air, water, moisture, or dust around the door or window sash.

Wood Rays—Strips of cells extending radially within a tree and varying in height from a few cells in some species to 4 inches or more in oak. The rays serve primarily to store food and to transport it horizontally in the tree.

Index

A

Air-and kiln-dried wood, 329
Air compressor, paint, 294
  accessories, 300
  air volume and pressure, 295
Aluminum siding, 107
Asphalt
  shingles, 29
  tile floor, 139
Attic and roof repair, 310
Automatic stroke belt sanders, 236

B

Band saws, 175-184
  circular arcs cutting, 179
  construction of, 175
  multiple-sawing operations, 179
  pointers on operation, 180
  ripping operations, 178
  straight-cutting operations, 177
Barytes, paint, 275
Basement
  maintenance, 307
  stairs, 118
Belt sanders, 234
Bench grinder, 256
Bevel siding, 99
Blue star, 330
Boring machines, 245-250
  tools, 246
Box cornice, 45

Brushes, paint, 290
Built-up roof, 21

C

Calipers
  hermaphrodite, 201
  inside, 198
  outside, 197
Canvas roofing, 19
Care of paint brushes, 291
Carriage, stair, 115
Casement windows, 88
Cement-asbestos shingles, 33
Centering and mounting stock,
  lathe, 199
Centers, lathe, 194
Ceramic floor tile, 141
Characteristics of woods for
  painting, 288
Circular
  arcs, bandsaw, 179
  saws, 165-174
    construction of, 165
    operations of
      crosscutting, 168
      grooving, 169
      mitering, 168
      ripping, 167
Closed cornice, 45
Coping, 60
Corner treatment, siding, 106
  metal corners, 107

Cornice
  construction, 45-51
    box, 45
    closed, 45
    open, 46
    rake or gable-end, 49
    wide box, 45
  returns, 47
Construction of stairs, 111
Crawl-space area, 309
Crosscutting operations, circular
  saw, 168
Cutting long miters, 59

D

Damage, termite, 265
Deadwood, 326
Density of wood, 322
Design of stairs, 113
Detecting termite damage, 268
Detection of roof leaks, 41
Disappearing stairs, 121
Disk sanders, 240
Doors, 65-83
  flush, 67
  frames, 70
  garage, 78
    fiberglass, 80
    steel, 80
    wood, 80
      openers, 82
    hanging mill-built, 73
  hollow-core, 68
  installation, 69
  jambs, 70
  louver, 68
  manufactured, 65
  mill-made, 65
  panel, 65
  sliding, 77
  solid-core, 67
  swinging, 76
  trim, 73
Double-hung windows, 85
Downspouts, 38
Driers, paint, 276
Drives, lathe, 195
Drop siding, 100
Drum sanders, 233
Dry-rot, 327
Dry-wall finish, 154

E

Electric
  drill, 253
  plane, 258
Estimating
  density of wood, 323
  paint needed, 292
Expanded metal lath, 146
Exterior
  painting, 282
  stairs, 121

F

Felt, roofing, 12
Fiberboard
  and wood interior walls, 161
  sheathing, 95
Finish moulding miter joints, 54
Finishing wood floors, 286
Flashing, roofing, 17
Floor
  covering, 131
    asphalt, 139
    ceramic tile, 141
    linoleum, 138
    rubber tile, 140
    wood, 137
  maintenance, 317
  sound-proof, 137
Flooring, 129-142
  porch, 141
  sub, 129
Flush doors, 67
Frames,
  door, 70
  window, 85
Framing of stairwell, 113

G

Garage doors, 78
  fiber glass, 80
  steel, 80
  wood, 80
    openers, 82
Glazing sash, 91
Glossary of
  housing terms, 333
  stair terms, 123-127
Grooving operations, circular saw,
  169

Gutters, 38

H

Hand
  -block sander, 242
  -lever stroke belt, 237
  saw, power, 251
Handrail, 118
Hanging mill-built doors, 73
Hardwood characteristic, 319
Hermaphrodite calipers, 201
Hinged or casement windows, 87
Hips, roofing, 27
Hollow-core doors, 68

I

Identification of termites, 264
Inside calipers, 198
Installation of
  doors, 69
  roof, 12
  siding, 103
Installing gypsum base, 147
Interior walls and ceiling, 143-163
  dry-wall, 154
  masonry, 158
  painting, 280
  plaster, 143
  plywood panels, 159
  wood and fiberboard, 161

J

Jambs, door, 70
Jigsaw, 185-190
  construction of, 185
  operation of, 188
Jointers, 213
  adjustments, 215
  construction of, 214
  operation of, 215

K

Knives, paint, 289

L

Lath walls, 143
  nailing, 150
Lathe

  attachments, 194
  speed, 192
Linoleum floor, 138
Linseed oil, 275
Louver doors, 68

M

Maintenance and repair, 307-318
Manufactured doors, 65
Masonry walls as plaster base, 158
Metal
  lath, 145
    plaster, 146
  siding, 107
Mill-made doors, 65
Miter work, 53-64
  coping, 60
  cutting long joints, 59
  finish moulding, 54
  tools, 53
Mitering operations, circular saw,
  168
Moisture content of wood, 320
Mortisers, 227
Multiple-sawing operations, band
  saw, 179

N

Nailing lath, 150
Newel and handrail, 118

O

Open cornice, 46
Outside calipers, 197

P

Paint mixtures, 276
Painting
  equipment, 273-306
  tools, 289
    brush, 290
    knives, 289
Paints, 273
  spray, 293
Paneled doors, 65
Paper, sheathing, 98
Physical characteristics of wood,
  319-331

Pigment, paint, 274
Placing the work, spray painting, 302
Planers, 225
    adjustments, 211
    construction of, 225
    operation of, 213
Plaster
    grounds, 148
    reinforcing, 143
Plastering material and method of application, 152
Plywood
    and wallboard painting, 287
    panels, 159
    sheathing, 96, 102
Pointers on band-saw operation, 180
Polishing, 204
Porch flooring, 141
Portable sander, 260
Power
    and speed,
        circular saw, 171
        shaper, 224
    operated hand tools, 251-261
        bench grinder, 256
        electric,
            drill, 253
            plane, 258
        hand saw, 251
        portable sander, 260
        saber saw, 259
Primary colors, 277
Priming coat, 281

R

Rake or gable-end finish, 49
Ratio of riser to tread, 112
Returns, cornice, 47
Ripping operations,
    band saw, 178
    circular saw, 167
Roll roofing, 11
Roof and attic repair, 310
Roofing, 9-43
    asphalt shingles, 29
    built-up, 21
    canvas, 19
    cement-asbestos shingles, 33
    detecting leaks, 41
    felt, 12

roll, 11
    selecting materials, 39
    sheet-metal, 36
    slate, 34
    wood shingles, 23
Rubber floor tile, 140

S

Saber saw, 259
Sanding machine, 233-245
    automatic stroke, 236
    belt, 234
    disk, 240
    drum, 233
    hand
        -block, 242
        -lever stroke belt, 237
    spindle, 242
    variety, 238
Sapwood versus heartwood, 330
Sash, window, 87
    glazing, 91
    installation, 87
    weights, 90
Screens, window, 92
Selecting roof materials, 34
Shapers, 218
    construction of, 218
    operation of, 219
Sheathing and siding, 95-109
    fiberboard, 95
    paper, 98
    plywood, 96, 102
    wood, 95
Sheet-metal roofing, 36
Shield, termite protection, 266
Shrinkage of wood, 321
Shutters, 93
Siding,
    aluminum, 107
    bevel, 99
    corner treatment, 107
    drop, 100
    installation of, 103
    metal, 107
    square-edge, 100
    treated, 103
    vertical, 101
    water table, 104
    wood, 98
Slate roof, 34

Sliding doors, 77
Slope of roof, 10
Soft wood characteristics, 319
Soil poison, termite protection, 269
Solid-core doors, 67
Sound-proof floors, 137
Spindle sanders, 242
Spray
    booth, 301
    guns, 298
    painting, 293
Spraying
    equipment connections, 300
    technique, 303
Square-edge siding, 100
Stairs, 111-128
    basement, 118
    carriage, 115
    construction of, 111
    design of, 113
    disappearing, 121
    exterior, 121
    framing of, 113
    glossy of, 123-127
    handrail, 118
    newel, 118
    ratio of riser and tread, 112
    stringers, 115
Starter course, roofing, 31
Starting and stopping the lathe, 193
Stopping termite damage, 269
Straight-cutting operations, band
        saw, 177
Strength of wood, 325
    compression, 325
    shear, 325
    tension, 325
Stringers, 115
Subflooring, 129
Sublimed white lead, paint, 274
Swinging doors, 76

T

Tenoners, 229
Termite protection, 263-271
    damage, 265
    identification of, 264
    shields, 266

soil treatment, 269
Time for cutting timber, 329
Tools,
    boring machine, 246
    miter, 53
Treated siding, 103
Treatment of hardened brushes, 291
Trim, doors, 73
Turpentine, 276
Type of nails, siding, 105
Typical woodturning operations, 202

V

Valleys, roof, 16, 28
Variety sanders, 238
Vehicles, paint, 275
Vertical siding, 101
Virgin and second growth, 327

W

Water table, siding, 104
White lead, paint, 274
Whiting, 275
Wide box cornice, 45
Windows, 85-98
    casement, 88
    double-hung, 85
    frames, 85
    hinged or casement, 87
    sash, 87
    screens, 92
    shutters, 93
Wood
    and fiberboard, 161
    sheathing, 95
    shingles, 23
    siding, 98
    strip flooring, 132
    -tile flooring, 137
Woodturning
    lathes, 191-205
    operations, 202
    tools, 196

Z

Zinc white, paint, 274